PROCEEDINGS
OF
THE JOHNS HOPKINS WORKSHOP
ON
CURRENT PROBLEMS IN PARTICLE THEORY

8

PREVIOUS JOHNS HOPKINS WORKSHOP PROCEEDINGS

1	BALTIMORE	'74	....
2	BALTIMORE	'78	....
3	FLORENCE	'79	....
4	BONN	'80	Lattice Gauge Theories. Integrable Systems.
5	BALTIMORE	'81	Unified Field Theories and Beyond.
6	FLORENCE	'82	Lattice Gauge Theories. Supersymmetry and Grand Unification.
7	BONN	'83	Lattice Gauge Theories. Supersymmetry and Grand Unification.

The Johns Hopkins Workshops on Current Problems in Particle Theory are organized by the following universities:

UNIVERSITY OF BONN

UNIVERSITY OF FLORENCE

THE JOHNS HOPKINS UNIVERSITY

ORGANIZING COMMITTEE

Gabor DOMOKOS (Johns Hopkins)
Susan KOVESI-DOMOKOS (Johns Hopkins)

Roberto CASALBUONI (Florence)
Luca LUSANNA (Florence)

Klaus DIETZ (Bonn)
Vladimir RITTENBERG (Bonn)

PROCEEDINGS OF THE JOHNS HOPKINS WORKSHOP
ON
CURRENT PROBLEMS IN PARTICLE THEORY 8

BALTIMORE, 1984
(June 20-22)

PARTICLES AND GRAVITY

Edited by
G. Domokos and S. Kovesi-Domokos

World Scientific

Published by
World Scientific Publishing Co Pte Ltd.
P O Box 128, Farrer Road, Singapore 9128

PROCEEDINGS OF THE 8TH JOHNS HOPKINS WORKSHOP ON CURRENT PROBLEMS IN PARTICLE THEORY — PARTICLES AND GRAVITY

ISBN 9971-966-90-5
9971-966-91-3 pbk

Printed in Singapore by Kim Hup Lee Printing Co Pte Ltd.

FOREWORD

This is the eighth in a series of Workshops on Current Problems in Particle Theory. As in the past, the basic purpose of this Workshop has been to provide a forum for theoretical physicists from all over the world to discuss important problems of current research in theoretical particle physics in an informal atmosphere. The discussions centered around a few invited talks, reproduced in this volume. The speakers summarized the current state of the art in their subfields and also presented new results.

This year the topic has been "Particles and Gravity", reflecting the exciting developments due, largely, to a cross-pollination between modern theoretical particle physics and cosmology. While "classical" cosmology as one knew it even a few years ago, relied upon the particle physics of the fifties (at best), it left several important observational facts, such as the observed isotropy of the 3°K background, the apparent spatial flatness of the Universe, etc. largely unexplained. Particle physicists, in the meantime, made predictions regarding the spectrum and interactions of "elementary" particles at energy scales of 10^{14} GeV or higher, without any immediate hope of testing those predictions by experiments done at accelerators as we know them today.

The "new" cosmology of the eighties holds the promise, on the one hand, of resolving the "puzzles" left behind by classical ("Gamowian") cosmology; on the other, it may provide the necessary observational constraints on theories of elementary particles in an energy region which, probably for a long time to come, will remain inaccessible to terrestrial experimentation.

It is too early to tell whether this promise will be fulfilled. However, this question can be resolved only by an active interaction between particle physicists and cosmologists, as it took place at this Workshop. It is our hope, therefore, that this volume will be useful in further stimulating one of the most exciting human endeavors: an ever deeper understanding of what our Universe is made of and how it developed until it reached its present shape.

We wish to thank our colleagues at the Department of Physics and Astronomy, Johns Hopkins University: G. Feldman, B. R. Judd (Chair) and C. W. Kim, for their enthusiasm, time and advice contributed during the period this Workshop was being organized and run; R. A. Zdanis, Vice Provost and G. W. Fisher, Dean, School of Arts and Sciences for their support of the Johns Hopkins Workshops. Without their help and support, this Workshop could not have succeeded in achieving its goals.

We also wish to express our thanks to Ms. Barbara Dreyfus for her dedicated and valuable help in organizing and running the Workshop and in producing the manuscript of these Proceedings.

We gratefully acknowledge financial support received from the U. S. Department of Energy, the National Science Foundation and the School of Arts and Sciences, Johns Hopkins University.

Last, but not least, we thank Dr. K. K. Phua, Editor-in-Chief, and the staff of World Scientific Publishing Co. for their understanding and encouragement in producing this volume.

The Organizing Committee

CONTENTS

PROCEEDINGS

OF

THE JOHNS HOPKINS WORKSHOP

ON

CURRENT PROBLEMS IN PARTICLE THEORY

8

THE PARTICLE ZOO: IT KEEPS GROWING

Hector Rubinstein

CERN
Theory Division
CH-1211 Geneva 23
SWITZERLAND

and

Weizmann Institute of Science†
Rehovot 76100
ISRAEL

In this lecture I would like to give a theoretical particle physicist's view of the present status of particle physics. This is hardly a theoretical review. It is mainly an analysis of what is known and where new results are pointing to new physics.

The standard model is $SU_{3\,colour} \times SU_2 \times U(1)$. There are many aspects of these models that require verification. However it was a remarkable situation as a whole that until the UA(i) revolution i=1,2 it looked unchallenged.

We first analyse SU_3 colour. The gauge (unbroken) theory of quarks and gluons has incredible structure buried in the deceptive Lagrangian

$$\mathcal{L}_o = \frac{1}{4} G^a_\mu G^a_{\nu\mu} + g \sum_i \bar{\psi} (D + m_i)\psi \ .$$

The parameters of the theory are Λ_{QCD} that set scales through dimensional transmutation and the masses of quarks which are outside the theory. More precisely they are not calculable in QCD itself. Though not proven, this Lagrangian produces only singlet colour states. There is good evidence that hard processes obey QCD laws. Here we want to emphasize results on the parameters of the theory.

---

†Permanent address.

Λ^{MS}_{QCD} seems to be settling at about 150 MeV with 30% error. It is reassuring that deep inelastic scattering, decay widths, and QCD spectroscopy all point to this value.

The gluon pole is seen in Rutherford scattering by UA1. As a consequence the $1/1\text{-}\cos\theta$ enhancement in the angular distribution in jet production is clearly seen. The data necessitate running α and their value is Λ_{MS} = 200 MeV also in agreement with other determinations. Another area of QCD physics that seems open to interesting developments is QCD bulk matter. If one has a large box of quark matter and these, say 1000 quarks, are bound, the Fermi momentum is lowered by allowing different species to appear by weak interaction decays. Since the pressure of one flavour goes like $P_F{}^4$, the common equilibrium pressure of three quark flavours is given by the Fermi momentum

$$\tilde{\mu} = \frac{1}{3}(1 + 2^{4/3})^{1/3} .$$

The average kinetic energy is then lowered by a factor $(3/1+2^{4/3})^{3/4}$ = .89, a net gain of the order of perhaps 100 MeV provided the bag pressure is about the same as in a small bag. Adding the mass of the strange quark at the scale 300 MeV (Fermi momentum), this mixture has less energy than normal matter: a fascinating possibility. These calculations are very preliminary but I am sure refinements will be made[4]. The masses of quarks is a more subtle question. Current algebra and PCAC give information on quark mass ratios[1] and QCD sum rules[2] have given (as expected) very low masses for u and d quark masses. $m_u + m_d \sim 10$ MeV. These parameters are difficult to pin down better, though their difference $m_d - m_u \simeq 3$ MeV is well established from ρ-ω interference and $\psi' \to \pi\psi$ decay. These isospin violating processes <u>and</u> the analysis of baryon electromagnetic and quark mass differences give strong support to this result[3].

The strange quark mass has recently been shown to be very sensitive to QCD spectroscopy and the new determination shows[5])

$$m_s = 100 \pm 10 \text{ MeV at 1 GeV scale} .$$

This result is much less than expected but still a little bit too large as far as baryon QCD spectroscopy seems to want[6]. Low values are of interest because of the recent ideas that strange matter might be marginally more stable than normal matter[4], as we already discussed.

The heavier flavours are well determined from QCD sum rules[2] but more interesting it has been shown that at least m_c must run as a function of scale, something new to my knowledge.

$$m_c = 1.27 \pm 0.02 \text{ GeV}$$

$$m_b = 4.25 \pm 0.05 \text{ GeV} .$$

We know also that the spectroscopic calculations work well and predict very accurately all mesons with $\ell=0$ and $\ell=1$ (except for the $I=J=0$ sector where the U(1) anomaly makes calculation impossible).

The results are all in good agreement with experiment. The number of condensates needed is small, (this is understood)[2], and in particular the two driving power corrections are very well known[2]

$$\frac{\beta}{g} < G^a_{\mu\nu} G^a_{\mu\nu} > = [(360 \pm 20) \text{ MeV}]^3$$

$$<\bar{q}q_s> = .9 <qq_u> = <\bar{q}q_d> (1\text{-}2.10^{-2}) = (100 \text{ MeV})^3$$

as predicted by current algebra and the masses as described above.

Some contradiction with bottonium is not serious since the calculation of Voloshin makes unwarranted assumptions[2].

The recent preliminary evidence at 3.51 GeV for the $^1P_1$ state of charmonium in exact agreement with theory is very reassuring[2]. So far so good. However, there are still some clouds in the sky. The question is very simple. Has one really seen the true degrees of freedom of QCD? All the states mentioned are common to the normal quark model $\bar{q}q$ and qqq. It is clear that gluons are, in this theory, true degrees of freedom. We know that they are there, the deep inelastic sum rules show that momentum is carried away by glue. The showcase should be glueballs. All methods: bags, lattice and QCD sum rules give in some channel at least glue ball resonances. Our bias sides with SVZ[7)] that the scalar glueballs are probably strongly affected by vacuum mixing. However the relatively low lying J=2 states should be seen. It is unfortunate that above 1 GeV, their signature is not unique and these particles can be confused (and mixed!) with normal I=0 mesons or their excitations. This controversy as to whether some states are glueballs or not will not die easily but will not be solved soon either.

More promising are hybrid states. In $\bar{q}q$ systems some quantum numbers cannot be built by $\bar{q}q$ alone. $J^{PC} = 1^{-+}$ is an example. As a consequence, seeing such a resonance could indicate $\bar{q}qG$ or $\bar{q}q$ many gluons states. Obviously $\bar{q}q\bar{q}q$ is also possible but the strong evidence against exotic states makes, if states of this nature are seen, such an interpretatiön unlikely. Using QCD sum rules and bags, these states seem to be unavoidable. For light quarks, a I=1 m ~ 1.3 GeV state should decay into $\rho\pi$ or $\pi\eta$. It is slightly worrying that such an object defies verification. It should be seen in $p\bar{p}$ annihilation since the $q\bar{q}$ and the gluon are there, but no evidence is forthcoming. A forward backward asymmetry in $\pi\eta$ due to A_2 and p-wave interference[9)] has been reported: it is a necessary but not a sufficient conditon. For heavy quarks, the predictions are similar[10)]. The present LEAR experiments should soon decide. It is unfortunate that for baryons all possible states can be reached by three quarks and therefore we are in a similar undecisive situation as with glueballs. There is also a small hope in the overabundance for bottonium of radial excitations. Perhaps this is an indication of extra degrees of freedom but nothing definite seems in sight. To conclude this part: SU_3 colour is fine. Most predictions seem correct and the expected pattern of states and their widths is in good qualitative and sometimes quantitative agreement with experiment. There is lack of spectroscopic evidence for the gluonic degrees of freedom. However, the situation is far from settled. There are promising avenues of reseach like strange matter. Finally there are still unanswered questions concerning the θ angle and axions. Though there are some ideas on how to make progress along these lines, these are hard questions.

$SU_2 \times U(1)$

The remarkable pattern of $SU_2 \times U(1)$ has been verified to good accuracy. We discuss all parameters as seen today.

A. Lepton masses and mixing angles.

Besides the well known charged lepton masses, all efforts to establish a non-vanishing neutrino mass of any kind have not been confirmed. Intensive efforts in the last ten years have failed to produce a convincing result. The methods and determinations are

1. End point measurements. These difficult experiments show

$$m_{\nu_e} < 50 \text{ ev}, \qquad m_{\nu_\mu} < 500 \text{ kev}, \qquad m_{\nu_\tau} < 160 \text{ MeV}.$$

There is a non zero result on $m_{\nu_e} = 30$ eV but the experiment is subject to large corrections and is not considered as hard evidence. Several new experiments along these lines are underway: $H^3 \to H_\varepsilon{}^3 + \nu + e$.

2. Oscillations. If neutrinos mix, they oscillate and therefore as a function of distance beams appear and disappear. Unfortunately these counting rates are also functions of reactor properties and again though positive results keep being announced, there is no convincing evidence. The latest claim is $\Delta m^2 = (.02\text{ev})^2$ $\theta = 12°$ if $\nu_e = \cos\theta\ \nu_1 + \sin\theta\nu_2$.

3. Double the beta decay is sensitive to a Majorana mass. These experiments seem to give extremely good bounds. However, there is a caveat. If neutrinos have different CP properties, they can cancel each other's contribution and a sizeable Δm as reported in oscillation experiments could exist. It is unlikely but not incompatible.

4. Neutrino decays. There is a lot of speculation on neutrino decay. Formation of galaxies and other phenomena could be explained by massive neutrinos. This is very hard to measure.

B. The last member of the first three generations has been found. Looking at $W \to tb$ and $pp \to tt$ + anything, UA1 identifies events with one hard electron or muon and two jets. These come from the two b quark decays. Though mass determination is still uncertain, one has $m_t = 40 \pm 10$ GeV. The leptonic branching ratio fits expectations.

C. The Kobayashi-Maskawa matrix is not complete. The pattern is suggestive: as masses increase going away from the diagonal matrix elements these decrease as $1 \to \varepsilon \to \varepsilon^2$. However as a model of CP violation, the KM model is in trouble. The ε'/ε experiments favour now small negative numbers. If confirmed, KM is out (it predicts positive values of .2). The values are still compatible with the superweak model.

In conclusion, there is little evidence to abandon the $SU_2 \times U_1$ classification with massless neutrinos. The situation on ν masses will improve in the near future, angles and phases show a pattern to be understood.

Boson masses

There has been great experimental and some theoretical improvement on the matter.

UA(1) and UA(2) has given the following:

$$m_W = 83.1 \pm 1.9 \pm 1.3 \quad 80.9 \pm 1.5 \pm 2.4 \quad (83.0^{+2.9}_{-2.7} \text{ theory})$$

$$m_Z = 92.7 \pm 1.7 \pm 1.4 \quad 95.6 \pm 1.5 \pm 2.9 \quad (93.8^{+2.4}_{-2.2} \text{ theory})$$

$$m_Z - m_W \text{ (independent of } \sin\theta_W) = 10.8 \pm 0.6$$

$$\sin^2\theta_W = 0.215 \pm 0.001 \text{ low energy data } 0.217 \pm 0.014 \; .$$

The quantum effects give about 1 GeV shift and as seen in the errors will not be pinned down in the present generation experiments. Widths and production rates given the geometry are given by UA2 as

σ x BR ($W \rightarrow e\nu$) 530 ± 100 ± 100pb (350 ± 70 expected)

σ x BR ($Z \rightarrow e\bar{e}$) 110 ± 40 ± 20pb (~55 expected)

Γ_Z(direct) ~ 6.5 GeV

which implies less than 24 neutrino species. Both groups see almost no event $W \rightarrow e\nu\gamma$ (UA2 one candidate, expected 0.1) and therefore W physics seems totally conventional.

El convidado de piedra

	$e^+e^-\gamma$	UA1 and UA2	$\mu^+\mu^-\gamma$
$m(\ell^+\ell^-\gamma)$	98.8±5	90.6±1.9	$88.4\pm^{46}_{13}$
$m(\ell^+\ell^-)$	42.7±2.4	50.4±1.7	$70.9\pm^{37}_{12}$
$m(e\gamma)_2$	4.6±1.0	9.1±0.3	5.3±0.3
$m(e\gamma)_{21}$	88.5±2.5	74.7±1.8	~59

These events signal a pattern that though preliminary seems very exciting. Probably the Z° is misbehaving. No conventional bremstrahlung explanation is probable both in terms of γ-energy and production angle. Probabilities of $Z \rightarrow ee\gamma$ branching of far above 10% cannot be explained. Many papers have explained these events in terms of excited leptons, new bosons and other departures of SU_2 x U_1. All of these models are unconvincing. The forthcoming runs that will bring 10 times the number of Z°'s should confirm or not these events. It is also clear that supersymmetry is not able to explain this anomaly. Any boson below Z would create problems with weak interaction phenomenology. If confirmed, these events signal some remarkable new physics that defy easy understanding.

The forbidden Zoo

There is a handful of events in both UA_1 and UA_2 that go beyond $SU_2 \times U_1$. In one group there are always the Z-like configurations. Essentially, these events have too many multijet events as compared to what naive estimates of QCD on jet abundance could predict. Since the total sample is so small there is little evidence that is hard. Nevertheless, these events are all of the form A + missing energy. A is a photon, a jet, or a multijet. Assuming the missing energy is a $Z \to \nu\nu$ decay, all events point out to a particle decaying into $\gamma + Z$ or, jet + Z with mass $\sim$ 160 GeV.

The other group sees similar things in jet-jet invariant mass and also funny effects with W but there is some conflict between the two groups. Their data seem to some extent incompatible. These events again require confirmation and therefore the run coming soon will be invaluable. Theoretically, these events have also generated intense activity. Supersymmetry can fit some of these events. However as stated earlier, it can not bring overall order to the whole Zoo. If some mono-jet events are supersymmetric then gluinos must have about 40 GeV. Other explanations naturally remark that $160 \cong 2\text{x}80 \cong 2\text{x}M_W$ and therefore the weak interaction bosons are the ground state of a growing family. The detailed predictions if all data is of unique origin seem in trouble.

Conclusions

1. QCD is in good health: parameters getting to agree in different processes. Gluon effects appear "as they should".

2. QCD spectroscopy is well established with QCD sum rules. However, there is a problem with unseen degrees of freedom (glueballs and hybrids).

3. Strange matter 1/3 u, 1/3 d, 1/3 s seems possibly more stable than ordinary matter. The cosmological implications are serious but the question not settled: affaire a suivre.

4. θ angle and axions still a problem.

5. $SU_2 \times U(1)$ also works too well M_W and M_Z as well as $\sin\theta_W$ all in agreement with theory. Quantum effects are not detectable yet. As a tool for new physics, only useful probably at LEP.

6. $m_{\nu_i} = 0$, no serious evidence against this.

7. 3 generations complete.

8. Kobayaski Maskawa model for CP not doing well.

9. GUTS in trouble

   (a) proton lifetime too long already for minimal SU_5.
   (b) UA(1) Desert UA(2) $\rightarrow$ jungle at 100 GeV!
   (c) $Z \rightarrow \gamma + \ell^+\ell^-$ too many!
   (d) Z + multijets too many!
   (e) monojets + missing energy : something new?
   (f) 160 GeV resonance?
   (g) $\gamma + \nu$'s?
   (h) W + jets?

References

1) H. Gasser and H. Leutwyler, Physics Reports, 87, 77 1982.

2) H. Reinders, H.R. Rubinstein and S. Yazaki, Physics Reports, to appear.

3) M.A. Shifman, A.I. Vainshtein, and V.I. Zacharov, Nuclear Physics, B147, 385, 1979.
N. Isgur, H.J. Lipkin, H.R. Rubinstein and A. Schwimer, Physics Lett., 89B, 79, 1979 and references therein.

4) E. Witten, Princeton preprint and references therein.

5) H. Reinders, H.R. Rubinstein, Physics Letters, in print.

6) See for example B. Ioffe and collaborators, ITEP preprints.

7) V.A. Novikov, M.A. Shifman, A.I. Vainshtein, V.I. Zacharov, Nuclear Physics, 165, 67, 1980.

8) T. Barnes and F. Close and F. de Viron, Nucl. Phys. B224, 241, 1983.

9) T. Barnes private communication.

10) J. Govaerts, H. Reinders, H.R. Rubinstein, J. Weyers, in preparation.

GAUGED N=8 SUPERGRAVITY IN d=11 DIMENSIONS

B. de Wit*

Institute for Theoretical Physics
State University of New York at Stony Brook
Stony Brook, New York 11794

ABSTRACT

The embedding of pure gauged N=8 supergravity in d=11 supergravity is discussed. Explicit results are presented for two SO(7) invariant field configurations. These configurations contain the supersymmetric SO(8) invariant solution and two SO(7) invariant solutions of d=11 supergravity. One corresponds to the parallelized S^7; the other solution is new.

It is well-known that N=8 supergravity in d=4 space-time dimensions can be obtained by a truncation of d=11 supergravity in which the fields do not depend on the extra 7 dimensions.[1)] The method of dimensional reduction has been a useful tool for unraveling the structure of the d=4 theory. Irrespective of that one may wish to attribute physical significance to the extra dimensions; this leads to a supersymmetric generalization of Kaluza-Klein theory. In that context a large variety of solutions of d=11 supergravity[2)] has been found. They are all of the Freund-Rubin type[3)] in which the 4-rank antisymmetric field strength F_{MNPQ} acquires a nonzero value if all indices are in the 4-dimensional subspace corresponding to ordinary space-time

$$F_{\mu\nu\rho\sigma}(x,y) = ief\ \varepsilon_{\mu\nu\rho\sigma} \ . \qquad (1)$$

Here x^μ denotes the coordinates of 4-dimensional space-time and y^m the extra 7 coordinates; f is just a parameter to be determined by the field equations, and e is the vierbein determinant. If we assume that the 4-dimensional space-time is maximally symmetric then F_{MNPQ} may still be finite if all indices are in the 7-dimensional subspace, but

---

*) On leave from NIKHEF-H, Amsterdam

the components with mixed indices must be zero. Furthermore it is customary to decompose the d=11 metric tensor as

$$g_{MN}(x,y) = \begin{bmatrix} g_{\mu\nu}(x) & 0 \\ 0 & g_{mn}(y) \end{bmatrix} . \tag{2}$$

There are only two fully supersymmetric compactifications of this type[4]. In the first solution (1) vanishes and the 7- and 4-dimensional subspaces are flat. This corresponds to a compactification of the extra dimensions to the torus T^7, and the truncation mentioned above where all fields are y-independent comprises just the massless fluctuations about this ground state. In the other supersymmetric solution[5], the 4-dimensional subspace is an anti-de Sitter space and the extra 7 dimensions are compactified to the sphere S^7. The corresponding Riemann tensors are

$$\begin{aligned} R_{mnpq} &= m_7{}^2(g_{mp}g_{nq} - g_{mq}g_{np}) \ , \\ R_{\mu\nu\rho\sigma} &= m_4{}^2(g_{\mu\rho}g_{\nu\sigma} - g_{\mu\sigma}g_{\nu\rho}) \ , \end{aligned} \tag{3}$$

where $m_4{}^2 = 4m_7{}^2$ and $f^2 = 18m_7{}^2$, with $|m_7|$ the inverse radius of S^7. Hence the anti-de Sitter ground state is fully supersymmetric and invariant under the SO(8) isometries of S^7. Furthermore, the full spectrum of small fluctuations about this ground state has been determined, and consists of one massless supermultiplet [4,5] and an infinite tower[6] of massive N=8 anti-de Sitter supermultiplets[7] (the decomposition into supermultiplets confirms that the ground state is supersymmetric). There is one d=4 supergravity theory with the same features, namely gauged N=8 supergravity[8]. This theory has a supersymmetric anti-de Sitter ground state which leaves the local SO(8) group invariant (there are other versions of gauged N=8 supergravity where the gauge group is not SO(8), but these do not have a supersymmetric solution[9]). Therefore it seems obvious that the S^7 compactification of d=11 supergravity must correspond to gauged N=8 supergravity coupled to an infinite tower of massive supermultiplets.

If the above reasoning is correct one should be able to write down a d=4 field theory of supergravity coupled to matter which is equivalent to the full d=11 theory. Subsequently the matter multiplets can be put to zero, after which one is left with pure gauged N=8 supergravity. It is thus possible to truncate d=11 supergravity directly to N=8 supergravity by suitably restricting the y-dependence of the d=11 fields. The x-dependence is then characterized in terms of a finite number of x-dependent functions which should correspond to the d=4 fields. While such a restriction was straightforward on T^7 where the fields were taken constant in y, it is extremely complicated on S^7. A complete solution for the latter case is in fact not known. However, further knowledge of this truncation is highly desirable. It would rigorously establish the complete relationship between the S^7 compactification and N=8 supergravity, which may shed more light on the remarkable symmetry structure of the latter. Moreover, since solutions of pure supergravity are in general also solutions of a theory of supergravity coupled to matter, one would be able to formulate all d=4 solutions of pure N=8 supergravity[10] as solutions of d=11 supergravity. Note that the reverse is not always true! Solutions of d=11 supergravity do not necessarily coincide with solutions of pure N=8 supergravity, just as solutions of matter coupled to supergravity are not always solutions of pure supergravity. In fact, most of the known d=11 solutions are not related to pure N=8 supergravity, and only 2 known solutions have been identified as solutions of pure N=8 supergravity, namely the round[5] and the parallelized[11] S^7. So far none of the other d=4 solutions has been found in d=11 dimensions, and a first search for two particular solutions was in fact unsuccessful[12].

The work presented here (done in collaboration with H. Nicolai) aims at establishing the connection between the d=4 fields of pure N=8 supergravity and the d=11 fields. This can first be done in restricted background configurations. So far we have succeeded in identifying the values of the d=4 fields corresponding to d=11 field configurations with SO(7) invariance that are constant in ordinary space-time. This enables a quantitative comparison between the d=4 and d=11 supersymmetry transformation rules, and it leads to (possibly new) d=11 solutions whenever the d=4 fields are at an extremum of the N=8 supergravity potential.

To explain how this is done, let us return to the analysis of the small fluctuations about some classical solution. The d=11 fields are then conveniently expanded in terms of a suitable set of eigenfunctions $Y^{(n)}(y)$ of the relevant mass operator associated with the background

$$\phi(x,y) = \sum_n \phi^{(n)}(x)\, Y^{(n)}(y). \tag{4}$$

At first instance one might expect that restricting the decomposition (4) to the modes on S^7 corresponding to the massless supermultiplet defines the truncation of d=11 supergravity to gauged N=8 supergravity. However, the y-dependence specified in (4) is not free of ambiguity because the functions $Y^{(n)}(y)$ are subject to y-dependent gauge transformations which are a subset of the gauge transformations of the full d=11 theory. Furthermore the proper identification of the d=4 fields at a finite distance away from the background involves nonlinear modifications, as has been pointed out in ref. 13. Instead of specifying the y-dependence of $\phi(x,y)$ one may as well be specifying the y-dependence of $f(\phi(x,y))$ where f is some unknown function such that $f(\phi(x,y)) \propto \phi(x,y)$ in the linear approximation. Therefore from the result of the linearized analysis one cannot infer anything about the full y-dependence of the fields.

There is no doubt that the above complications play a role, because even in the simplest case of the T^7 reduction nonlinear field redefinitions and gauge choices are required for a proper identification of the d=4 fields. Moreover, one can show directly that the supersymmetry transformation rules for the modes in (4) are not consistent upon truncation to the massless sector, in the sense that modes belonging to massive supermultiplets transform into modes belonging to the massless supermultiplet. Therefore the massive supermultiplets cannot be put equal to zero, because they reappear through the supersymmetry transformations. In deriving this result one uses supersymmetry parameters that leave the S^7 background invariant; the y-dependence of these parameters is restricted by the Killing condition

$$\mathring{D}_m \mathring{\varepsilon}(x,y) = -\frac{1}{2} m_7 i \Gamma_m \mathring{\varepsilon}(x,y) \quad . \tag{5}$$

To make the transformation rules consistent one must exploit nonlinear field redefinitions of the type mentioned above. Also the supersymmetry parameters $\mathring{\varepsilon}(x,y)$ are then modified accordingly. All modifications vanish, however, in the S^7 background. In ref. 14 it has been demonstrated that redefinitions of this type are sufficient to make the linearized transformations consistent upon truncation to the massless multiplet.

Our strategy is to identify the fields of pure N=8 supergravity in d=11 supergravity by requiring that the field transformations are consistent upon truncation to the massless supermultiplet. To do this it is important that we first bring the transformation rules in a convenient form. Therefore we redefine the fields and the transformation parameters in such a way that the d=11 transformation rules resemble as closely as possible the transformation rules of pure N=8 supergravity. Somewhat surprisingly, this can be done without specifying either the background or the y-dependence of the fields[15]. Consequently these redefinitions are a direct generalization of the procedure followed for T^7 in ref. 1.

We do not present the explicit transformation rules here, but refer the reader to ref. 15. The important observation is that only a restricted set of (nonlinear) field redefinitions is allowed if one wishes to preserve the qualitative features of these transformation rules (linear redefinitions are not interesting in this context). It turns out that those redefinitions must take the form of field-dependent chiral SU(8) rotations. Therefore field-dependent chiral SU(8) rotations are the only possibility for making the supersymmetry transformations consistent upon truncation to the massless supermultiplet. In the S^7 solution the SU(8) rotations are zero, and when the fields move away from this solution one expects the fields to rotate under SU(8). Since the rotation depends on the fields, which in turn depend on y, the y-dependence of the modes thus changes continuously and differs from the y-dependence that was previously found for the small fluctuations about the supersymmetric solution.

This approach has now been used to analyse two types of SO(7) invariant field configurations. These configurations are characterized by two real 4-rank antisymmetric tensors C^{+}_{IJKL} and C^{-}_{IJKL}. Both are selfdual with respect to the 8-index Levi-Civita symbol with opposite duality phase. Therefore they transform according to two inequivalent 35-dimensional representations of SO(8). Furthermore, C^{+}_{IJKL} and C^{-}_{IJKL} leave an SO(7) subgroup invariant, but because they belong to inequivalent SO(8) representations, the corresponding invariant SO(7) subgroups are also inequivalent. The deviation of the fields from their values on S^7 are expressed in terms of two y-dependent tensors

$$S_{abc} \propto \bar{\eta}^I(y)\Gamma_{[ab}\eta^J(y)\ \bar{\eta}^K(y)\Gamma_{c]}\eta^L(y)\ C^{+}_{IJKL} \quad , \qquad (6)$$

$$\xi_a \propto \bar{\eta}^I(y)\Gamma_{ab}\eta^J(y)\ \bar{\eta}^K(y)\Gamma^b\eta^L(y)\ C^{-}_{IJKL} \quad , \qquad (7)$$

where $\eta^I(y)$ are the 8 independent Killing spinors (i.e. the linearly independent solutions of (5) labelled by I=1,..,8), and Γ_a the Γ-matrices in 7 dimensions (a=1,...,7). Antisymmetrised products of Γ-matrices are denoted by Γ_{ab}, Γ_{abc}, etc. The choice of (6) and (7) may look arbitrary to the reader, but these quantities characterize some of the small fluctuations corresponding to the massless supermultiplet about the S^7 background. Therefore for small fluctuations (6) and (7) are expected to be directly related to the fields of pure N=8 supergravity, although it is a priori not guaranteed that this relationship will hold for finite deviations.

Subsequently one defines two field configurations, one in terms of (6) and the other in terms of (7) (field configurations depending on both (6) and (7) are not considered because their symmetry is at most G_2). The precise form of these field configurations is found by requiring that the supersymmetry variations only contribute to the modes belonging to the massless supermultiplet, after allowing for a field-dependent SU(8) redefinition. This guarantees that the massive supermultiplets can consistently be put equal to zero, and identifies the two field configurations as SO(7) invariant field configurations of pure N=8 supergravity in which the scalar and pseudoscalar fields take constant (i.e. x-independent) values. The d=11 field configurations that we find depend each on one parameter, just as the SO(7) invariant configurations in d=4, which is a first indication that our approach is correct.

The situation is shown schematically in Fig. 1, where we sketch the two one-parameter field configurations which intercept at the S^7 background. It is known that for a particular point of the C^+ configuration there is a solution of the d=11 field equations. This is the parallelized sphere[11], for which there are two degenerate solutions that differ by the overall sign of the relevant order parameter. In N=8 supergravity there is precisely such a degenerate solution if the pseudoscalar fields acquire an SO(7) invariant vacuum expectation value[10,13] (the degeneracy is due to symmetry under parity reversal). For the C^- configuration no solutions have been found so far, but it is known that N=8 supergravity has another SO(7) invariant solution where the scalar fields acquire a vacuum expectation value[10,13]; so we expect to find a solution somewhere on the C^- configuration as well.

To discuss some quantitative features of these two field configurations we give some definitions. The deviation of the siebenbein $e_m{}^a(x,y)$ from the S^7 background value $\overset{\circ}{e}_m{}^a(y)$ is parametrized by a matrix S according to

$$e_m{}^a(x,y) = \overset{\circ}{e}_m{}^b(y) \quad S_b{}^a(y) \ , \tag{8}$$

where a,b,... are 7-dimensional tangent-space indices. Since we assume a maximally d=4 background the right-hand side of (8) does not depend on x^μ. According to the field redefinitions that we have been discussing before the d=4 space is described by a vierbein field $e'_\mu{}^\alpha(x)$ which is related to the corresponding components $e_\mu{}^\alpha(x,y)$ of the d=11

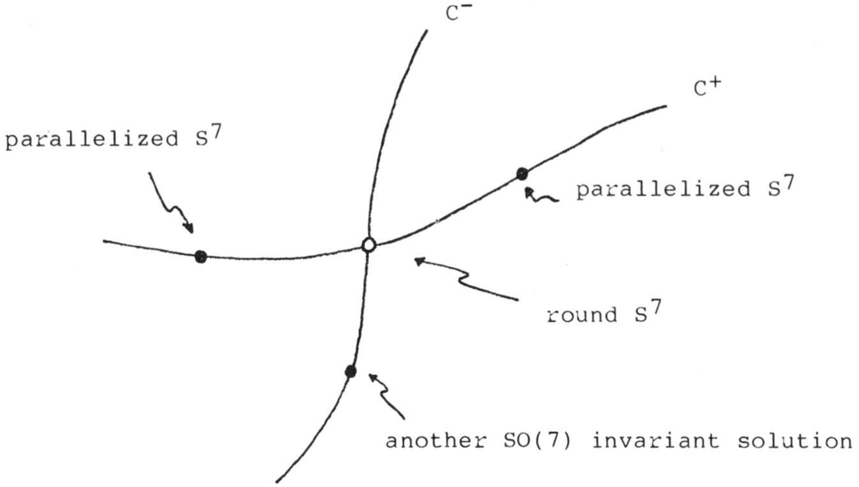

Fig. 1: Two SO(7) invariant field configurations in d=11 supergravity

elfbein by a scale transformation

$$e_{\mu}{}^{\alpha}(x,y) = \Delta^{-1/2}(y)e'_{\mu}{}^{\alpha}(x) \ , \tag{9}$$

where

$$\Delta(y) = \det(S(y)) \ . \tag{10}$$

Furthermore the nonzero components of the field-strength tensor F_{MNPQ} are expressed in terms of

$$f = -\frac{1}{24} i \ \Delta^{-2}(y)\varepsilon^{\alpha\beta\gamma\delta}F_{\alpha\beta\gamma\delta}(x,y) \ , \tag{11}$$

$$F^{abc}(y) = \frac{1}{24} \eta' \ \Delta^{-1/2}(y)\varepsilon^{abcdefg}F_{defg}(x,y), \tag{12}$$

where $F_{\alpha\beta\gamma\delta}$ and F_{abcd} are the purely 4- and 7-dimensional components of F_{ABCD}, respectively. Here A,B,... are d=11 tangent-space indices and $\alpha, \beta, \ldots$ belong to the d=4 subset. In a maximally symmetric d=4 background f does not depend on x^{μ} and y^m by virtue of the Bianchi identities for the field strength F_{MNPQ}; F^{abc} depends only on y.

Both field configurations contain the round S^7 background configuration corresponding to

$$S_{ab} = \delta_{ab} \,,$$
$$F_{abc} = 0 \,, \tag{13}$$
$$f = 3\sqrt{2}\, m_7 \,,$$

and the deviations of (13) are characterized in terms of the tensors (6) or (7).

In the C^+ configuration S^{ab} is y-independent and proportional to the unit matrix; F^{abc} is given by

$$F_{abc}(y) = 2\sqrt{2}\, m_7\, tg(4\tau_+)\, S_{abc}(y) \,, \tag{14}$$

where S_{abc} depends on y as indicated by (6) and τ_+ is an arbitrary parameter. The tensor S_{abc} is the so-called Cartan-Schouten torsion[17] that parallelizes S^7. Its defining properties are

$$\mathring{D}_m S_{npq} = \frac{1}{6} m_7\, \eta' \varepsilon_{mnpqrst} S^{rst} \,, \tag{15}$$

$$S^{[mnp} S^{q]rs} = -\frac{1}{4} \eta' \varepsilon^{mnpq[r}{}_{tu} S^{s]tu} \,, \tag{16}$$

$$S^{mnp} S_{qrp} = 2\delta^{mn}{}_{qr} - \frac{1}{6} \eta' \varepsilon^{mn}{}_{qrstu} S^{stu} \,, \tag{17}$$

where $\eta' = \pm 1$. The first equation allows 35 independent solutions; the other equations (16) and (17) restrict this to one solution modulo an overall SO(8) rotation. In (15)-(17) we have converted tangent-space indices into S^7 indices by contraction with the siebenbein $\mathring{e}_m{}^a(y)$ of S^7. For small values of the parameter τ_+ (14) defines one of the small fluctuations about S^7 that belong to the massless N=8 supermultiplet[5,6].

If (14) is inserted into the d=11 supersymmetry transformations one finds that the result is only consistent for the massless supermultiplet if one performs a uniform chiral SU(8) transformation

$$U(y,\tau_+) = \exp(\tfrac{1}{6}\tau_+ S^{abc}(y)\,\Gamma_{abc}\gamma_5) \,, \tag{18}$$

and the parameter f takes the value

$$f(\tau_+) = \sqrt{2}\, m_7 (3 - 4\, tg^2(4\tau_+))\, \Delta^{-3/2} \,. \tag{19}$$

To show the consistency is a rather tedious calculation, but on the basis of this result one may conclude that the C^+

configuration defined by (14) and (19) corresponds to a field configuration in pure N=8 supergravity in d=11 dimensions. Since the d=4 subspace was taken maximally symmetric and the chiral transformation (18) induces explicit γ_5 terms one can deduce that the corresponding N=8 configuration must arise from giving the pseudoscalar fields an SO(7) invariant vacuum expectation value. It is known that for one particular value of this vacuum expectation value there is an extremum of the N=8 theory. Likewise d=11 supergravity has a solution in the C^+ configuration namely at $|\text{tg } 4\tau| = 1/2$. This is the parallelized S^7 solution[11]. It is thus possible to compare the supersymmetry transformation rules of d=11 supergravity at the parallelized sphere to those of d=4 supergravity at the pseudoscalar solution. This has been presented in ref. 15, where a complete numerical agreement was found (we recall that both d=4 and d=11 solutions are degenerate; see Fig. 1).

As we have been emphasizing before the y-dependence of the fields that correspond to the massless supermultiplet of N=8 supergravity is changed by a y-dependent SU(8) transformation when one moves away from the S^7 background Therefore the small fluctuations about the parallelized S^7 that correspond to N=8 supergravity have a completely different dependence on y than the small fluctuations about the round S^7 that correspond to N=8 supergravity. This has been confirmed by an explicit calculation in which the spectrum of modes at the parallelized S^7 was compared to the spectrum at the pseudoscalar critical point of N=8 supergravity[18].

One may attempt to extend the results of the C^+ configuration by allowing all 35 independent solutions of (15). It turns out, however, that this background is not consistent with a field configuration in pure N=8 supergravity[15]. In other words one must conclude that also the massive supermultiplets have acquired nonzero values in this case. Undoubtedly this is related to the fact that the chiral SU(8) rotation mixes the modes associated with F_{abc} and S_{ab}, so that one must allow for a nontrivial value for $S_{ab}(y)$ as well. The reason that the C^+ configuration does correspond to pure N=8 supergravity must then be related to the fact that it is SO(7) invariant. This is suggested by the fact that no SO(7) invariant background exists in N=8 supergravity in which both scalar and pseudoscalars take nonzero values.

The C^- configuration is more complicated, and we restrict ourselves to the main results. In this case F_{abc}

vanishes, and the siebenbein depends on the vector ξ_a defined in (7). If we define $\hat{\xi}_a$ as the unit vector, and introduce another y-dependent quantity ξ related to its length as

$$\xi = \frac{1}{16}\, \bar{\eta}^I(y)\, \Gamma^a \eta^J(y)\, \bar{\eta}^K(y)\, \Gamma_a \eta^L(y)\, C^-_{IJKL} \quad , \qquad (20)$$

we can write $S_{ab}(y)$ as follows

$$S_{ab} = g^{-1/3}(\tau_-, \xi)\, \{\delta_{ab} + (\frac{g(\tau_-,\xi)}{1+21\tau_-} - 1)\, \hat{\xi}_a \hat{\xi}_b\} \, , \qquad (21)$$

with

$$g^2(\tau_-, \xi) = 1 + 63\tau_-^2 - 2\tau_-(1+g\tau_-)\xi \quad . \qquad (22)$$

The explicit form of (21) and (22) follows from requiring consistency upon trunction to the massless supermultiplet. The parameter τ_- is left undetermined by this condition, so we are again dealing with a one-parameter configuration. For small values of τ_- one finds

$$S^{ab} = (1-3\tau_-)\,\delta^{ab} + \frac{3\tau_-}{8}\, \{\bar{\eta}^I \Gamma^a \eta^J\, \bar{\eta}^K \Gamma^b \eta^L - \frac{1}{9}\,\delta^{ab} \bar{\eta}^I \Gamma^c \eta^J \bar{\eta}^K \Gamma_c \eta^L\}\, C^-_{IJKL} + \mathcal{O}(\tau^2) \, , \qquad (23)$$

which coincides with the small fluctuations about the S^7 background[5,6]. To obtain consistency with the massless supermultiplet an SU(8) field redefinition is again required which belongs to the SO(8) subgroup in this case. It can be expressed by

$$U^2(y,\tau_-) = g^{-1}(y,\tau_-)\, \{(1-\tau_-\xi)\mathbb{1} + \tau_-\sqrt{(21+\xi)(3-\xi)}\; i\Gamma^a \hat{\xi}_a\}. \qquad (24)$$

The parameter f is equal to

$$f(\tau_-) = \sqrt{2}\, m_7(1-3\tau_-). \qquad (25)$$

The most conspicuous aspect of the C^- configuration is that Δ is now y-dependent. Therefore, d=11 solutions in this configuration do not satisfy (2), but

$$g_{MN}(x,y) = \begin{pmatrix} \Delta^{-1}(y) g_{\mu\nu}(x) & 0 \\ 0 & g_{mn}(y) \end{pmatrix} . \qquad (26)$$

On the basis of d=4 arguments in N=8 supergravity we have already argued above that the C^- configuration must contain a solution of d=11 supergravity. We have explicitly verified by inserting (21) as an ansatz in the d=11 field equation that this is indeed the case[16)] Unlike solutions of type (2) the d=11 Riemann curvature tensor now also has nonvanishing components with two indices in the d=4 and two indices in the d=7 subspace. On the basis of these results we expect all but two of the N=8 supergravity solutions to have a d=11 counterpart of type (26).

References

1. E. Cremmer and B. Julia, Phys. Lett. 80B (1978) 48, Nucl. Phys. B159 (1979) 141.

2. E. Cremmer, B. Julia and J. Scherk, Phys. Lett. 76B (1978) 409.

3. P.G.O. Freund and M.A. Rubin, Phys. Lett. 97B (1980) 233.

4. B. Biran, F. Englert, B. de Wit and H. Nicolai, Phys. Lett. 124B (1983) 45.

5. M.J. Duff and C.N. Pope, in "Supersymmetry and Supergravity '82", eds. S. Ferrara, J.G. Taylor and P. van Nieuwenhuizen (World Scientific).

6. B. Biran, A. Casher, F. Englert, M. Rooman and P. Spindel, Phys. Lett 134B (1984) 179; A. Casher, F. Englert, M. Rooman and H. Nicolai, preprint TH3794-CERN; E. Sezgin, Trieste preprint IC-83-220.

7. H. Nicolai and D.Z. Freedman, Nucl. Phys. B237 (1984) 342.

8. B. de Wit and H. Nicolai, Phys. Lett. 108B (1981) 285, Nucl. Phys. B208 (1982) 323.

9. C. M. Hull, preprints MIT.

10. N. P. Warner, Phys. Lett. 128B (1983) 169.

11. F. Englert, Phys. Let. 119B (1982) 339.

12. M. Günaydin and N. P. Warner, preprint CALT-68-1077.

13. B. de Wit and H. Nicolai, Nucl. Phys. B231 (1984) 506.

14. M.A. Awada, B.E.W. Nilsson and C.N. Pope, Phys. Rev. D29 (1984) 334.

15. B. de Wit and H. Nicolai, Nucl. Phys. B, to appear.

16. B. de Wit and H. Nicolai, in preparation.

17. E. Cartan and J.A. Schouten, Proc. K. Akad. Wet. Amsterdam 29(1926)933.

18. B. Biran and Ph. Spindel, preprint ULB-TH 84/002.

REMARKS ON A POSSIBLE EFFECTIVE HADRONIC SUPERSYMMETRY*

Feza Gürsey

Yale University
Physics Department
Gibbs Laboratory
New Haven, CT 06511

ABSTRACT

It is shown that inside rotationally excited baryons, QCD leads to the formation of diquarks well separated from the remaining quark. At this separation the scalar, spin independent, confining part of the effective QCD potential is dominant. Since QCD forces are also flavor independent, the force between the quark q and the diquark (qq) inside an excited baryon is essentially the same as the one between q and the antiquark $\bar{q}$ inside an excited meson. Thus, the approximate spin-flavor independence of hadronic physics expressed by SU(6) symmetry gets extended to the Miyazawa U(6/21) supersymmetry through a symmetry between $\bar{q}$ and (qq), resulting into the parallelism of mesonic and baryonic Regge trajectories. Various aspects and implications of this approximate effective supersymmetry and its breaking are discussed.

1. Introduction

Supersymmetry is a symmetry between fermions of half odd integer spin and bosons of integer spin. In supergravity, for example, we have an action that is invariant under operations transforming fields of spins 2, 3/2, 1, 1/2, and 0 among themselves[1)]. Here we recall that low lying hadrons have the same range of spin values with s=0,1 and 2 mesons interlaced with s=1/2 and 3/2 baryons. Groups that transform s=0 and 1/2

---

*Research supported in part by the U.S. Department of Energy Contract No. DE-AC-02-76ER0 3075.

mesons into each other and simultaneously mix s=1/2 and 3/2 baryons have been proposed long ago[2]. This kind of symmetry had to be a broken symmetry since no degeneracy is observed between particles of different spin. However, a cursory glance at the Chew-Frautschi plot of hadronic Regge trajectories[3] reveals the following features:

a) All trajectories with j (hadronic spin) versus m^2 (m = hadronic mass) are approximately linear.

b) Leading mesonic trajectories associated with lowest spin 0, 1 and 2 are parallel among themselves.

c) Leading baryonic trajectories associated with s=1/2, 3/2 are also parallel among themselves.

d) Mesonic and baryonic trajectories are approximately parallel to each other with a universal hadronic slope $\alpha' \approx 0.9\ (\mathrm{GeV})^{-2}$.

e) The separation between mesonic trajectories is nearly the same as the one between baryonic trajectories.

Now, properties (b) and (c) suggest the existence of a phenomenological symmetry between mesons of different spin, which also operates on baryons with different spin. For hadrons composed of the light quarks u, d, s, this symmetry is expressed by the group SU(6) x O(3) where O(3) describes the rotational excitations on the leading trajectories and the spin-flavor group SU(6) classifies the lowest elements of the trajectories into particle multiplets. The property (d), on the other hand, tells us that there is a new kind of symmetry (supersymmetry) between the bosonic mesons and the fermionic baryons. The universal Regge slope for hadrons is then a supersymmetric observable.

The meaning of the property (e) is that, the physical mechanism that breaks the SU(6) symmetry must also be responsible for the breaking of its supersymmetric extension. Finally, from property (a) we infer that the potential binding the quarks is approximately linear and we have to apply relativistic quantum mechanics appropriate to light quarks while the Schrödinger nonrelativistic theory is sufficient for the description of quarkonium systems for heavy quarks.

According to the quark model of Gell-Mann and Zweig, mesons and baryons are described respectively by bound ($\bar{q}q$) and (qqq) systems. Any symmetry between mesons and baryons must correspond at the quark level to a supersymmetry between $\bar{q}$ (antiquarks) and bound (qq) states (diquarks). Now $\bar{q}$, with $s=1/2$ and unitary spin associated with the triplet (3) representation of the flavor SU(3) belongs to the (6) representation of SU(6). The low lying baryons are in its symmetric (56) representation. Since (56) is contained in $6 \times 21 = 56 + 70$, the diquark with $s=0$ or 1 must be in the symmetric (21) representation of SU(6). Hence, the hadronic supersymmetry we are seeking must transform the ($\bar{6}$) and (21) SU(6) multiplets, both color antitriplets into each other and therefore must be 27 dimensional with 6 fermionic and 21 bosonic states. Such a supergroup which is now called U(6/21) was introduced by Miyazawa[4)] in 1967 as a generalization of the hadronic SU(6) symmetry, following earlier attempts in this direction by Hwa and Nuyts[5)] and also himself[6)]. On the strength of this work Miyazawa can be regarded as the inventor of super-Lie algebras that form a closed system under commutators as well as anticommutators and satisfy a generalized Jacobi identity that is explicitly worked out for supergroups U(m/n) in Ref. 4.

Miyazawa also realized that the ($\bar{q}$)-(qq) symmetry that also implies a (q)-($\bar{q}\bar{q}$) symmetry will transform the meson ($q\bar{q}$) in general not only to baryons (qqq) and antibaryons ($\bar{q}\bar{q}\bar{q}$) but also to exotic mesons ($\bar{q}\bar{q}$)(qq) that belong to the SU(6) representations 1, 35 and 405. The (1) and (35) are 0^+ and 1^+ mesons while the (405) also includes mesons with spin 2^+ and isospin 2. All the low lying hadrons will now be in the adjoint representation of U(6/21) with both spin and isospin taking values 0, 1/2, 1, 3/2, and 2.

The next introduction of supersymmetry into Physics was within the context of dual resonance models that evolved later into string models. These theories due to P. Ramond[7)] and Neveu and Schwarz[8)] lead naturally to linear baryon and meson trajectories that are parallel. The string models were not local and they were not relativistic in four dimensions. It was difficult to apply them to Particle Physics, although they are now making a comeback in fundamental Physics as a relativistic finite super-symmetric Kaluza-Klein type superstring theory in 10 dimensions[9]. Examples of renormalizable relativistic local quantum field theories involving s=0, 1/2 and 1 fields with interaction were first constructed by Zumino and Wess[10] following initial attempts by other authors[11]. Wess and Zumino based their work on the super-Poincaré algebra that is a supersymmetric generalization of the infinitesimal Poincaré group and sits in the superconformal algebra. The final step of incorporating s=2 and s=3/2 fields in a local relativistic field theory was taken by the discoverers of supergravity[1] and extended supergravity who were able to supersymmetrize Einstein's general relativity and Kaluza-Klein theories. This work culminated into the N=8 supergravity in 4 dimensions or the N=1 supergravity in 11 dimensions[12] which is a theory that is not renor-malizable, but may be finite and incorporates internal symmetry as well as fields with spins ranging from 0 to 2. All these local supersymmetric quantum field theories have fascinating convergence and symmetry proper-ties but they are far from describing properties of hadrons or even quarks and leptons. If these fundamental fields have any physical reality at all, they may be associated with preons (or haplons) that would be hypothetical constituents of quarks, leptons and fundamental gauge bosons.

Returning to the more concrete world of hadrons, we may try to see if the phenomenological approximate group SU(6) and its supersymmetric extension by Miyazawa can be justified within the standard theory of colored quarks interacting through gluons that are associated with the color gauge group $SU(3)^c$. The justification of SU(6) and the derivation of its breaking was given by Georgi, Glashow and de Rujulà[13].

On the other hand, a phenomenological supersymmetry in Nuclear Physics between odd and even nuclei was discovered and formulated by means of supergroups[14] U(m/n) by Balantekin, Bars and Iachello[15]. It was natural to see if such groups could also describe hadronic supersym-metry. It was in this context that Balantekin, Bars and Tze[16] rediscovered Miyazawa's U(6/21) supersymmetry which was called V(6/21) by Miyazawa[4]. This group commutes with the color group and acts on $\bar{q}$ and (qq) both color antitriplets.

In this talk I propose to extend the approach of Georgi et al.[14] to see how far it can provide a basis for the existence of an approximate hadronic Miyazawa supersymmetry. My remarks are based on joint work with Sultan Catto[17]. It turns out that most of the ingredients for this approach are already in the literature.

The key concepts are:

a) The approximate validity of the string theory as an approximation to QCD (Quantum Chromodynamics). This was shown by the lattice gauge theory as a strong coupling approximation to QCD, following the pioneering work of Wilson[18] and also by the elongated bag model of Johnson and Thorn[19] following the bag model[20] approximation to QCD.

b) The emergence of the diquark structure in QCD through the string approximation. This was done by Eguchi[21] and also by Johnson and Thorn[19].

c) The vector (spin dependent) nature of the Coulomb part together with the scalar (spin independent) nature of the confining part of the q - $\bar{q}$ potential. These properties were collectively worked out by many authors using both perturbation theory and lattice gauge theory methods[22].

d) The necessity for introducing exotic ($\bar{q}\bar{q}$) (qq) mesons in QCD, through the bag approximation[23], the string picture[24] or the confining potential model[25].

e) Deviations from linearity of Regge trajectories within the context of QCD[3].

Once all the parts of this jigsaw puzzle are put together, a rather simple picture emerges. The diquark behaves very much like the antiquark in the strong coupling regime because QCD forces are flavor independent and the confining part of the QCD potential is spin independent. This immediately leads to a $\bar{q}$ - (qq) effective supersymmetry at large separation. At short distances, spin is approximately conserved because of asymptotic freedom but spin independence gets broken through one-gluon exchange in the perturbation theory regime. The diquark structure also disappears at short distance, leading to the breaking of both SU(6) and U(6/21) at low energies. There is another difficulty associated with a supersymmetric extension of SU(6) noted by Salam and Strathdee[26]. Because of the anticommutativity of Grassmann numbers and also of supercharges, antisymmetrical representations of SU(6) like (15) for diquarks and (20) for baryons will occur in hadronic supermultiplets. An example is N=8 supergravity where only antisymmetrical representations of O(8) occur. In the case of colored quarks, however, it is possible to introduce non associative Grassmann numbers u_1, u_2 and u_3 constructed out of octonion units[27]. They transform like a triplet under the SU(3) subgroup of the automorphism group G_2 of octonions. Then the octonionic quarks $\vec{u} \cdot \vec{q}_\alpha$ and $\vec{u} \cdot \vec{q}_\beta$ will commute unlike $q^i{}_\alpha$ and $q^j{}_\beta$ that are anticommutative (α and β are combined spin-flavor indices). This procedure which is the basis of color algebra[28] not only converts antisymmetrical spin-flavor group representations into symmetrical ones, but as shown by Domokos and Kövesi-Domokos, suppresses the color sextet (6) representations for diquarks that would otherwise be allowed under the rule for the preservation of the Pauli principle for the colored quark states. With the introduction of octonionic quarks we have shown that quarks, diquarks, mesons and baryons can be all viewed as elements of an octonionic superalgebra[17].

The simplest supersymmetric Hamiltonian is obtained starting from semi-relativistic dynamical model of quarks and diquarks already used by Lichtenberg et al.[29] for an approximate calculation of baryonic masses. It is possible to use a spin representation of the Wess-Zumino algebra to write first order relativistic equations for quarks and diquarks that are invariant under supersymmetry transformations[30]. Concluding remarks will deal with such Dirac-like supersymmetric equations and with a brief discussion of experimental possibilities for the observation of the diquark structure and exotic $(\bar{q}\bar{q})$ (qq) mesons.

2. The diquark and semi-relativistic Hamiltonians for (qq) and q-(qq) systems

In QCD, both q-$\bar{q}$ and q-q forces are attractive, the color factor of the latter being half that of the former. Thus, the formation of diquarks inside a baryon is a definite possibility. Various strong coupling approximations to QCD, like lattice gauge theory[18,22], 't Hooft's 1/N approximation (when N, the number of colors is very large), or the elongated bag model[19] all give a linear potential between widely separated quarks and an effective string that approximates the gluon flux tube. In such a theory, Eguchi[21] has shown that it is energetically favorable for the three quarks in a baryon to form a linear structure with a quark in the middle and two at the ends or, for high rotational excitation, a bilocal linear structure with one quark at one end and a diquark at the other end. This was reconfirmed by Johnson and Thorn independently in the bag model when the bag is deformed and elongated in a rotationally excited baryon[19]. Thus if we move along a leading baryon Regge trajectory to a region of high j, we are likely to find the baryon as a bilocal object consisting of a quark and a diquark instead of a trilocal object that represents a ground state baryon more accurately. On the other hand a meson is also a bilocal object consisting of a quark and an antiquark interacting via a linear potential at large separation. Consider then a bilocal object with constituents with respective masses m_1, m_2, spins $\vec{s}_1$, $\vec{s}_2$ and color representations 3 and $\bar{3}$. One of the constituents can be the quark q or the anti-diquark $\bar{D}$ (both color triplets) while the other can be an antiquark $\bar{q}$ or a diquark D (qq) which are both color antitriplets. The QCD force between the two constitutents will be flavor independent and it will consist of two parts: a Coulomb like part V_c which transforms like the time component of a 4-vector (due to the exchange of single s=1 gluons at short separations) and a confining part V_s which is largely a relativistic scalar, is spin independent and is due to the exchange of a great many gluons that form a flux tube or an elongated bag at large separation of the constituents.

Let $\vec{p}_1 = \vec{p}$ and $\vec{p}_2 = -\vec{p}$ be the center of mass momenta of the two constituents. The quantity which is canonically conjugate to the relative coordinate $\vec{r} = \vec{r}_1 - \vec{r}_2$ is $-i\vec{\nabla} = -i\ \partial/\partial\vec{r} = 1/2(\vec{p}_1 - \vec{p}_2) = \vec{p}$. Ignoring the center of mass motion, following Lichtenberg et al.[30], we can write a semi-relativistic wave equation for the wave function $\Psi_{12}(\vec{r})$ of the bilocal object with energy eigenvalues W_{12}, namely,

$$(W_{12} - V_c)\Psi_{12} = \left[(m_1 + \frac{1}{2}V_s)^2 - \nabla^2\right]^{1/2} + \left[(m_2 + \frac{1}{2}V_s)^2 - \nabla^2)\right]^{1/2} \Psi_{12}. \tag{2.1}$$

The scalar and vector potentials are given by

$$V_s = br \quad , \quad V_c = -\frac{4}{3}\frac{\alpha_s}{r} + k_{12}\ \frac{\vec{s}_1\cdot\vec{s}_2}{m_1 m_2} \tag{2.2}$$

where 4/3 is the color factor, α_s is the strong coupling constant at the energy W_{12}, and the spin dependent part of the vector potential is the hyperfine structure correction due to gluon exchange with $k_{12} = |\Psi_{12}(0)|^2$. We see that at large r, neglecting the mass difference $(m_2 - m_1)$, we find the same equation for both the $(q\text{-}\bar{q})$ and the q-D system, except for the presence of the hyperfine term that breaks the symmetry between $\bar{q}$ and D. To this approximation, we can transform the second constituent $\bar{q}$ into D and vice-versa without changing the energy eigenvalue W. This means that the system admits the approximate U(6/21) supersymetry transformation

$$\begin{aligned} \delta\bar{q}^i_\alpha &= \bar{b}_{\alpha\beta\gamma}(D^i)^{\beta\gamma} \\ \delta(D^i)^{\beta\gamma} &= b^{\alpha\beta\gamma}\ \bar{q}^i_\alpha \end{aligned} \tag{2.2}$$

in addition to the SU(6) transformation

$$\delta\bar{q}^{i}_{\alpha} = m^{\beta}_{\alpha}\,\bar{q}^{i}_{\beta} \qquad (2.3)$$

$$\delta(D^{i})^{\beta\gamma} = n^{\beta\gamma}_{\rho\sigma}\,(D^{i})^{\rho\sigma} .$$

The breaking of both SU(6) and U(6/21) is due to the hyperfine term while the supersymmetry is further broken by the quark-diquark mass difference $m_2 - m_1$.

We could also have brought into play the wave functions Ψ_D and $\Psi_{\bar{q}}$ of the diquark and the antiquark at point $\vec{r}_2$ in the field of the quark at point $\vec{r}_2$. The masses m_1 and m_2 must then be replaced by the reduced masses

$$\mu_D = \frac{m_q m_D}{m_q + m_D} \quad , \quad \mu_{\bar{q}} = \frac{m_q m_{\bar{q}}}{m_q + m_{\bar{q}}} = \frac{1}{2} m_q \, . \qquad (2.4)$$

In this case, the wave function belongs to the fundamental 27 dimensional representation of U(6/21) and the Hamiltonian commutes with the supersymmetry transformation (2.2) of the wave function except for the difference in the reduced masses μ_D and $\mu_{\dot{q}}$ and the hyperfine structure term in the Hamiltonian.

The breaking of supersymmetry results in the mass difference

$$\Delta m \cong m_2 - m_1 + k\,\frac{\vec{s}_1 \cdot \vec{s}_2}{m_1 m_2} \qquad (2.5)$$

at high energies both for baryons and mesons. At low energies, the baryon becomes a trilocal object with 3 quarks and the mass splitting is given by[14])

$$\Delta m_{123} = \frac{1}{2} k \left(\frac{\vec{s}_1 \cdot \vec{s}_2}{m_1 m_2} + \frac{\vec{s}_2 \cdot \vec{s}_3}{m_2 m_3} + \frac{\vec{s}_3 \cdot \vec{s}_1}{m_3 m_1} \right) \qquad (2.6)$$

where m_1, m_2 and m_3 are the masses of the three different quark constituents.

Going back to the first formulation which is more symmetrical, we can write the Hamiltonian associated with Eq. (2.1) in a U(6/21) covariant form: we have four cases for Ψ_{12}, namely $\Psi_{q\bar{q}}$, $\Psi_{\bar{D}D}$, Ψ_{qD} and $\Psi_{\bar{D}\bar{q}}$ that represent mesons, exotic mesons, baryons and antibaryons with respective dimensions 6x6, 21x21, 6x21, and 21x6 that all fit into the following 27x27 adjoint representation of U(6/21):

$$\Psi(r) = \begin{bmatrix} \Psi_{q\bar{q}}(r) & \Psi_{qD}(r) \\ \Psi_{\bar{D}\bar{q}}(r) & \Psi_{\bar{D}D}(r) \end{bmatrix} . \tag{2.7}$$

The wave equation for the hadronic wave function Ψ can now be written as

$$i \frac{\partial \Psi}{\partial t} = H\Psi = \{K, \Psi\} + \vec{S}\, \Psi\, \vec{S} . \tag{2.8}$$

where K is the diagonal matrix

$$K = \begin{bmatrix} K_q I^{(6)} & 0 \\ 0 & K_D I^{(21)} \end{bmatrix} , \quad I^{(n)} = \text{nxn unit matrix}, \tag{2.9}$$

with

$$\begin{aligned} K_q &= -\frac{2}{3}\frac{\alpha_s}{r} + \left[(m_q + \frac{1}{2} V_s)^2 - \nabla^2\right]^{1/2} , \\ K_D &= -\frac{2}{3}\frac{\alpha_s}{r} + \left[(m_D + \frac{1}{2} V_s)^2 - \nabla^2\right]^{1/2} \end{aligned} \tag{2.10}$$

and

$$\vec{\xi} = k^{1/2} \begin{bmatrix} m_q^{-1}\vec{\xi}_q I^{(6)} & 0 \\ 0 & m_D^{-1}\vec{\xi}_D I^{(21)} \end{bmatrix} .$$

The second, spin dependent term on the r.h.s. of Eq. (2.8) is a symmetry breaking term for both U(6) and U(6/21). The first term is U(6) symmetrical. It also preserves the supersymmetry in the limit of $m_q = m_D$. Hence if the quark-diquark mass difference and the spin dependent terms are neglected, Eq. (2.8) is invariant under the U(6/21) infinitesimal transformation

$$\delta\Psi = [N, \Psi] \tag{2.11}$$

where

$$N = \begin{bmatrix} M & b \\ \bar{b} & N \end{bmatrix} \tag{2.12}$$

where M and N have elements $m_\alpha{}^\beta$ and $n_{\rho\sigma}{}^{\beta\gamma}$ respectively in Eq. (2.3) while the rectangular matrices b and $\bar{b}$ have elements $b^{\alpha\beta\gamma}$ and $\bar{b}_{\alpha\beta\gamma}$ occuring in Eq. (2.2). In that limit, the Hamiltonian H commutes with the generators of the transformation (2.11).

3. <u>The hadronic Regge trajectories</u>

Consider the Hamiltonian of Eq. (2.1). We can write

$$-\nabla^2 = \vec{p}^2 = p_r^2 + \frac{\ell(\ell+1)}{r^2} \tag{3.1}$$

where ℓ is associated with the orbital excitation of the system. For high rotational excitations, the expectation value of r is large, corresponding to a stretched string. The angular momentum ℓ is also large. The value of the centrifugal energy which is proportional to $\ell(\ell+1)/r^2$ has a similarly large value. Since V_s which is proportional to r will also have a high absolute value, the constituent masses become negligible in the high relativistic limit. On the other hand, the radial excitation term p_r^2 can be neglected on the leading trajectory associated with the lowest radial energy. Consequently the Hamiltonian can be approximated by the expression

$$h = 2\left[\frac{b^2r^2}{4} + \frac{\ell(\ell+1)}{r^2}\right]^{1/2} = \left[b^2r^2 + \frac{4\ell(\ell+1)}{r^2}\right]^{1/2} . \quad (3.2)$$

The expectation value of r will correspond to the length of the stretched rotating string which is given by the value $r = r_0$ that minimizes h, i.e.

$$r_o^2 = \frac{2\sqrt{\ell(\ell+1)}}{b} . \quad (3.3)$$

Replacing r by its expectation value r_0, we obtain

$$h \approx h(r_o) = \sqrt{2}\, br_o = 2\sqrt{b}\, [\ell(\ell+1)]^{1/4} . \quad (3.4)$$

The squared mass of the system is given by h^2 and its spin $\vec{j}$ by $\vec{\ell} + \vec{s}$ where $\vec{s} = \vec{s}_1 + \vec{s}_2$. Hence

$$j(j+1) = \ell(\ell+1) + s(s+1) + 2\vec{\ell}\cdot\vec{s} . \quad (3.5)$$

For large ℓ we can replace ℓ by j and find the relation

$$m^2 = 4b\sqrt{j(j+1)} \cong 4bj \quad (3.6)$$

showing that the leading trajectories are asymptotically linear with slope $\alpha' = 1/(4b)$. A calculation that takes into account the relativistic contribution of the string gives $\alpha' = 1/(2\pi b)$ [31]. Starting from $b = 0.18 - 0.2\ (\mathrm{GeV})^2$ from lattice QCD or the bag theory, one finds[19] $\alpha' = 0.88\ (\mathrm{GeV})^{-2}$, in excellent agreement with the experimentally given universal slope.

A better approximation including the V_c term yields

$$[h(r_o) - k\,\frac{\vec{s}_1\cdot\vec{s}_2}{m_1 m_2}\,]^2 \;=\; \frac{(m_1+m_2)^2}{2} + 2(m_1^2 + m_2^2 - \frac{4}{3}\,\alpha_s\, b\sqrt{2}\,)$$

$$+ 4b\,\sqrt{\ell(\ell+1)} + 2(m_1+m_2)\,\sqrt{2b}\,[\ell(\ell+1)]^{1/4} + O(\ell^{-1/2})\;. \qquad (3.7)$$

This expression determines the deviation of the Regge trajectory from linearity as well as the values of the Regge intercepts for different constituents. It is known[3] that hadronic trajectories are not exactly linear and curve downwards from their asymptotic straight lines for lower values of j. The equation (3.7) provides a better approximation to the experimental trajectories than the linear trajectories given by the string theories.

For $\ell=0$ and p_r negligible, we find

$$H \;=\; m_1 + m_2 + V_s + V_c \qquad (3.8)$$

which is the usual sum of the rest mass energy and the QCD potential energy.

4. The hadronic color algebra

We can now put the color singlet hadrons in Eq. (2.7) together with quarks and diquarks by making use of the octonionic split units defined by[27]

$$u_j = \frac{1}{2}\,(e_j + ie_{j+3})\;,\quad u_j^* = \frac{1}{2}(e_j - ie_{j+3})\;,\quad (j = 1,2,3), \qquad (4.1a)$$

$$u_o = \frac{1}{2}\,(1 + ie_7)\;,\qquad u_o^* = \frac{1}{2}(1 - ie_7)\;. \qquad (4.1b)$$

They obey the $SU(3)^c$ invariant relations

$$u_i u_j = \varepsilon_{ijk} u_k^*\;,\quad u_i u_j^* = -u_o \delta_{ij}\;, \qquad (4.2a)$$

$$u_o^2 = u_o \ , \quad u_o u_j = u_j u_o^* = u_j \ , \quad u_o^* u_j = u_j u_o = 0 \qquad (4.2b)$$

and their complex conjugate relations. The U_j can be regarded as exceptional Grassmann numbers since they satisfy

$$\{u_i, u_j\} = 0 \quad , \quad \{u_i, u_j^*\} = -\delta_{ij} \qquad (4.3)$$

but they are not associative.

We now introduce the octonionic valued quark and diquark fields by

$$Q_A = u_i Q_A^i = \vec{u} \cdot \vec{Q}_A \ , \quad D_{AB} = D_{BA} = Q_A Q_B = \vec{u}^* \cdot \vec{D}_{AB} \ ,$$

$$\bar{Q}_A = u_i^* \bar{Q}_A^i = \vec{u}^* \cdot \vec{Q}_A^* \ , \quad D_{AB}^* = D_{BA}^* = \bar{Q}_A \bar{Q}_B = \vec{u} \cdot \vec{D}_{AB}^* \ . \qquad (4.4)$$

We can construct the color singlet baryonic fields

$$Q_A(Q_B Q_C) = - u_o B_{ABC} \qquad (4.5)$$

in the (56) symmetrical representation. Properties of this algebra are discussed by Domokos and Kövesi-Domokos[28]. Consider the 28x28 octonionic matrix

$$Z = \begin{bmatrix} u_o M & u_o B & \vec{u}\cdot\vec{Q} \\ u_o B^\dagger & u_o N & \vec{u}\cdot\vec{D}^* \\ \varepsilon\vec{u}^*\cdot\vec{Q}^\dagger & \varepsilon\vec{u}^*\cdot\vec{D}^T & u_o^* L \end{bmatrix} . \qquad (4.6)$$

Here ε can be given values 1, -1 or 0. M and N are respectively 6x6 and 21x21 hermitian matrices, B a rectangular 6x21 matrix $\vec{u} \cdot \vec{Q}$ a 6x1 column matrix, $\vec{u} \cdot \vec{D}^*$ a 21x1 column matrix and L a 1x1 scalar. Such matrices close under anticommutator operations for $\varepsilon=1$. Matrices iZ close under commutator operations. In either case, they do not satisfy the Jacobi identity. But for $\varepsilon=0$, when the algebra is no longer semi-simple, the Jacobi identity is satisfied and we obtain a hadronic superalgebra which is an extension of the algebra U(6/21). Its automorphism group includes $SU(6) \times SU(21) \times SU(3)^c$. Thus color is automatically incorporated.

5. Relativistic formulation

There is a spin realization of the Wess-Zumino Super-Poincaré algebra

$$[P_\mu, P_\nu] = 0 , \quad [D_\alpha, P_\mu] = 0 , \quad [\bar{D}_{\dot\beta}, P_\mu] = 0 , \qquad (5.1)$$
$$[D^\alpha, \bar{D}^{\dot\beta}] = \sigma_\mu^{\alpha\dot\beta} P^\mu$$

with P_μ transforming like a 4-vector and D^α, $\bar{D}^{\dot\beta}$ respectively like left handed and right handed spinors under the Lorentz group with generators $J_{\mu\nu}$. The finite non unitary spin realization is in terms of 4x4 matrices for $J_{\mu\nu}$ and P_ν

$$J_{\mu\nu} = \frac{1}{2}\sigma_{\mu\nu} = \frac{1}{4i}[\gamma_\mu, \gamma_\nu] , \quad P_\mu = \pi_\mu^L = \frac{1-\gamma_5}{2}\gamma_\mu . \qquad (5.2)$$

Introducing two Grassmann numbers θ_α ($\alpha = 1,2$) that transform like the components of a left handed spinor and commute with the Dirac matrices γ_μ, we have the representation

$$D_\alpha = \Delta^\alpha = \frac{\partial}{\partial\theta_\alpha} , \quad \bar{D}^{\dot\beta} = \bar{\Delta}^{\dot\beta} = \theta_\alpha \sigma_\mu^{\alpha\dot\beta}\pi_\mu^L . \qquad (5.3)$$

Such a representation of the Super-Poincaré algebra acts on a Majorana chiral superfield

$$X(x, \theta) = \psi(x) + \theta_\alpha C^\alpha(x) + \frac{1}{2}\theta_\alpha\theta^\alpha \phi(x) . \qquad (5.4)$$

Here ψ and ϕ are Majorana superfields associated with fermions and C^α has an unwritten Majorana index and a chiral spinor index α, so that it represents a boson.

On the other hand, we have the realization of P_μ in terms of the differential operator $-i\partial_\mu = -i\partial/\partial x^\mu$. In the Majorana representation, the operator $\gamma_\mu\partial_\mu$ is real and $\psi^c = \psi^*$.

Let us define ψ_L and ψ_R by

$$\psi_L = \frac{1}{2}(1 + \gamma_5)\,\psi\ , \quad \psi_R = \frac{1}{2}(1 - \gamma_5)\psi = \psi_L^* \ . \tag{5.5}$$

The free particle equation can now be written

$$\pi^L_\mu \partial_\mu \psi_L = m\psi_L^* \ . \tag{5.6}$$

We can introduce

$$X_L = \frac{1}{2}(1 + \gamma_5)X\ , \quad X_R = \frac{1}{2}(1-\gamma_5)X = X_L^* \ . \tag{5.7}$$

Then, eq. (5.6) generalizes to the superfield equation

$$\pi^L_\mu\, \partial_\mu\, X_L = mX_L^* \ . \tag{5.8}$$

Now consider the supersymmetry transformation

$$\delta X_L = (\xi^\alpha \Delta_\alpha + \bar{\xi}_\beta \bar{\Delta}^\beta)X_L = \Omega\, X_L \ . \tag{5.9}$$

This transformation commutes with the operator $\pi^L_\mu \partial_\mu$ so that

$$\pi^L_\mu \partial_\mu (X_L + \delta X_L) = m\,(X_L + \delta X_L)^* \ . \tag{5.10}$$

If ψ_L is a left handed quark and $C^\alpha(x)$ an antidiquark with the same mass as the quark, Eq. (5.9) provides a relativistic form of the quark anti diquark symmetry which is in fact broken by the quark-diquark mass difference. The scalar supersymmetric potential is introduced through $m \to m + V_S$ as before and the equation (5.8) remains supersymmetric. By means of this formalism, it is possible to reformulate the treatment of section 2 in first order relativistic form.

6. Discussion and Concluding Remarks

The success of Miyazawa's approximate hadronic symmetry depends on the existence of diquarks inside excited baryons and the occurrence of

$(\bar{q}\bar{q})(qq)$ exotic mesons in the hadronic spectrum. The diquark structure in the baryon will cause the form factors in lepton baryon scattering to deviate from the model with three point-quarks in the bag. Donachie et al.[32] and Fredrickson et al.[33] have shown that the inclusion of both spin one and spin zero diquark states in the nucleon can explain some of the deviations from the quark-parton model without including higher order gluon corrections. According to our analysis such diquark states should occur in deep inelastic scattering only at high impact parameter when the nucleon is thrown into a rotationally excited state.

The exotic meson states should occur whether one starts from a string model[24] a relativistic oscillator model[25] or a bag model[23] within the mass range 1.2 - 2 GeV. Their experimental absence is a real puzzle. However, recently some possible evidence for the formation of s=2, I=2 meson resonances that would fit nicely in the (405) representation of SU(6) has been found in the analysis of $\gamma\gamma \to \rho\rho$ reactions by Li and Liu[34]. More experimental confirmation must be forthcoming before the issue is settled.

To conclude, we may say that no supersymmetric extension of QCD seems to be necessary to explain the supersymmetry implied by the universal slope of hadronic trajectories. A rotationally excited hadron becomes a bilocal object with the well separated two constituents interacting through a linear spin independent and flavor independent potential. The constituents can be $q - \bar{q}$, $q - D$, $\bar{q} - \bar{D}$ or $D - \bar{D}$ and there is a natural supersymmetry between $\bar{q}$ and D or q and $\bar{D}$. Supersymmetry is broken by the mass difference of the constituents. Both supersymmetry and the phenomenological SU(6) symmetry are broken by spin dependent interactions induced by single gluon exchanges.

Acknowledgments

Besides my collaborator Sultan Catto, I would like to thank I. Bars, C. Tze, F. Iachello, G. Domokos, S. Kövesi-Domokos and Y. Hosotani for their interest and their valuable guidance to publications that deserve to be better known.

References

1) D.Z. Freedman, P. von Nieuwenhuizen and S. Ferrara, Phys. Rev. D13, 3214 (1976),
S. Deser and B. Zumino, Phys. Lett. 62B, 335 (1976).

2) F. Gürsey and L. Radicati, Phys. Rev. Lett. 13, 173 (1964).

3) For the present status of Regge trajectories, see A. Hendry, Phys. Rev. Lett. 41, 222 (1978) and A.J. Hey and R.L. Kelly, Phys. Reports 96, 71 (1983).

4) H. Miyazawa, Phys. Rev. 170, 1586 (1968).

5) R.C. Hwa and J. Nuyts, Phys. Rev. 151, 1215 (1966).

6) H. Miyazawa, Prog. Theor. Phys. (Kyoto) 36, 1266 (1966).

7) P. Ramond, Phys. Rev. D3, 2415 (1971).

8) A. Neveu and J. Schwarz, Nucl. Phys. B31, 86 (1971).

9) M. Green and J. Schwarz, Nucl. Phys. B223, 125 (1983).

10) J. Wess and B. Zumino, Nucl. Phys. B70, 39 (1974).

11) Y.A. Golfand and E.P. Likhtman, JETP Lett. 13, 323 (1971),
D.V. Volkov and V.P. Akulov, Phys. Lett. 46B, 109 (1973).

12) E. Cremmer and B. Julia, Nucl. Phys. B159, 141 (1979),
B. de Wit and H. Nicolai, Nucl. Phys. B231, 506 (1984).

13) A. de Rujulà, H. Georgi and S.L. Glashow, Phys. Rev. D12, 147 (1975).

14) For the classification of supergroups, including U(m/n), see V.G. Kac, Comm. Math. Phys. 53, 31 (1977).

15) A.B. Balantekin, I. Bars and F. Iachello, Phys. Rev. Lett. 47, 19 (1981).

16) A.B. Balantekin, I. Bars and H.C. Tze (private communication).

17) S. Catto and F. Gürsey, "Algebraic Treatment of Effective Supersymmetry", Yale preprint YTP84-07 (1984).

18) K.G. Wilson, Phys. Rev. D10, 2445 (1974).

19) K. Johnson and C.B. Thorn, Phys. Rev. 13, 1934 (1976).

20) A. Chodos, R.L. Jaffe, K. Johnson and C.B. Thorn, Phys. Rev. D10, 2599 (1974).

21) T. Eguchi, Phys. Lett. 59B, 475 (1975).

22) For a recent computation, see S.W. Otto and J.D. Stark, Phys. Rev. Lett. 52, 2328 (1984).

23) R.L. Jaffe, Phys. Rev. D15, 281 (1977).

24) H. Harari, Phys. Rev. Lett. 22, 562 (1969),
J.L. Rosner, Phys. Rev. Lett. 22, 689 (1969).

25) P.J. Mulders, A.T. Aerts and J.J. de Swart, Phys. Rev. D21, 2653 (1980).

26) A. Salam and J. Strathdee, Nucl. Phys. B87, 85 (1975).

27) M. Günaydin and F. Gürsey, J. Math. Phys. 14, 1651 (1973).

28) G. Domokos and S. Kövesi-Domokos, J. Math. Phys. 19, 1477 (1978).

29) D.B. Lichtenberg, W. Namgung, E. Predazzi and J.G. Wills, Phys. Rev. Lett. 48, 1653 (1982).

30) S. Catto and F. Gürsey, to be published.

31) E. Eichten et al. Phys. Rev. D17, 3090 (1978), Phys. Rev. D21, 203 (1980),
C. Quigg and J.L. Rosner, Phys. Lett. 71B, 153 (1977).

32) A. Donachie and P.V. Landshoff, Phys. Lett. 95B, 437 (1980).

33) S. Fredricksson, M. Jändel and T. Larsson, Z. Phys. C19, 53 (1983), Phys. Rev. Lett. 51, 2179 (1983).

34) B.A. Li and K.F. Liu, Phys. Rev. Lett. 51, 1510 (1983).

A RELATION BETWEEN N=2 SUPERGRAVITIES IN 10 AND 4 DIMENSIONS*

G. Chapline**
Lawrence Livermore National Laboratory
Livermore, California 94550

and

R. Slansky***
Theoretical Division
Los Alamos National Laboratory
Los Alamos, NM 87545

ABSTRACT

We list the 4-dimensional supergravities that are derived from the N=2, 10-dimensional supergravities using the method of coset-space dimensional reduction. The spectrum of each 4-dimensional theory (Table I) consists of those supermultiplets that are invariant under the symmetry transformations of the coset space.

The N=2 supergravity theories in D=10 dimensions[1)] are of special importance because they are anomaly free[2)] and because they can be obtained as a certain limit of superstrings[3)]. There are just two representations of D=10, N=2 supersymmetry that contain a single spin 2 graviton in D=4; each representation classifies the spectrum of a D=10, N=2 supergravity theory. These N=2 supermultiplets are direct sums of two N=1 representions, one being the N=1 gravitational multiplet.

The physical degrees of freedom are classified by representations of the SO(8) subgroup of the D=10 light-cone symmetry. For example, the gravitational N=1 supermultiplet has the SO(8) content, $\underline{1} + \underline{8}_s + \underline{28} + \underline{35}_v + \underline{56}_s$. The D=4 helicity content is obtained from the branching rules of these representations into representations of SO(6) X U(1), where the helicity is the eigenvalue of the U(1) generator. Table II lists the necessary branching rules; the conventions of Ref. 4) are followed. The graviton is in the $\underline{35}_v$, and the 4 helicity 3/2 states are in the $\underline{56}_s$ and transform as $\overline{\underline{4}}$ of the SO(6).

The N=2 multiplets are completed with one or the other of the two N=1 supermultiplets with largest helicity 3/2.

The SO(8) content of the N=1, spin 3/2 supermultiplet for the type A theory is $\underline{8}_v + \underline{8}_c + \underline{56}_v + \underline{56}_c$; the helicity 3/2 states transform as a $\underline{4}$ of the SO(6). The 8 gravitinos of the N=2, type A theory are self conjugate under SO(6) ($\underline{\bar{4}} + \underline{4}$), since the type A theory can be derived from D=11 simple supergravity. The type B theory is a direct sum of the N=1 gravitational multiplet and the other N=1, spin 3/2 multiplet, which is composed of $\underline{1} + \underline{8}_s + \underline{28} + \underline{35}_c + \underline{56}_s$; the 8 gravitinos transform as $\underline{\bar{4}} + \underline{\bar{4}}$ under the SO(6).

Our purpose here is to list the D=4 supergravities that are derived from these 10-dimensional theories by coset-space dimensional reduction[5,6]. The results are summarized in Table I. There are 5 spaces for which the D=4 theory possesses a supersymmetry with N<8. We do not list cases yielding D=4, N=8 supergravity, for then, the method becomes that of Cremmer and Julia[7]. Also, many reduced theories have no supersymmetry and no fermions at all, and are not included in Table I. The precise count of scalar and vector supermultiplets depends on the coset space used to describe the 6 extra dimensions. The first 4 cases in Table I give D=4, N=2 supergravities, and the last case reduces to N=4 supergravity coupled to 4 N=4 vector multiplets. The type B theories differ from type A in the first three cases; type A and B are equivalent in the last 2 cases. Although there is a hint of chirality in that types A and B sometimes differ, neither is flavor-chiral theory.

In most D>4 theories, the geometry of the vacuum is assumed to be some 4-dimensional space-time (such as Minkowski or anti de Sitter) times a compact space, usually a coset space S/R, where S is the isometry group of the extra dimensions and R is the stability group. If the extra dimensions are physical, then the 10-dimensional fields can be expanded into an infinite tower of 4-dimensional fields, where the D=4 fields are the coefficients of S-representation functions in the harmonic expansions of the 10-dimensional fields on S/R[8]. One way to simplify such a complicated theory is to truncate the harmonic expansions to include only a finite number of D=4 fields. However, to be useful, the Lagrangian for this subset of D=4 fields should be an effective low-energy theory for the D=10 theory, and the vacuum for this truncated theory should be a stable solution to the whole theory. (We take the extra dimensions seriously, and not merely as a formal device to obtain a 4-dimensional theory.)

One approach would be to keep only the zero-mass fields. However, in practice one is usually limited to looking for zero modes of the theory after it has been linearized about a vacuum solution[9]; then it is often unclear what the effective low-energy theory really is[10].

Also, the zero modes of the linearized theory may result because of a cancellation between a large negative contribution to the mass from anti de Sitter space and a large positive contribution from the curvature of S/R[9]. In D=10 supergravities the anti de Sitter background can be avoided if the D=10 scalar fields in the theory acquire a mass[11].

The coset-space reduction is a very different truncation of the theory. This involves keeping only the S-invariant fields. An advantage of this approach is that, if the truncated theory has a vacuum state, then the S-symmetric fields are guaranteed to be a solution to the full equations of motion[12]. Note that the S-symmetric vector fields are not contained in the D=10, N=1 gravitational supermultiplet in the first 4 cases of Table I. Thus, we lose the attractive geometrical interpretation of the vector particles of the Kaluza-Klein reduction as gauge symmetries of S, but it will be interesting to learn whether or not the S-symmetric vectors gauge some other symmetry.

The existence of this relationship between N=2 supergravity theories in 10 and 4 dimensions may be interesting for other reasons. The structure of matter and gauge field couplings to N=2 supergravity in 4-dimensions has been the subject of several recent papers[13]. Although the Lagrangians describing the interactions of the supermultiplets with N=2 supergravity are not derived here, the Lagrangian derived from Type A supergravity with N=2 scalar multiplets only will presumably have the form described by Bagger and Witten, where the scalar fields parameterize a quaternionic space.

The mathematical problem of coset-space dimensional reduction is to find just those fields that are invariant under the transformations on S. We follow Manton's light-cone prescription for coset-space dimensional reduction[5]. Thus, the dimensionally reduced theory is a truncation of the harmonic expansions on S/R to S singlets. With this observation in mind, it is trivial to extend the analysis to include cases were S/R is a group manifold or a product of a group manifold and a coset space. For example, if the space is a 6-dimensional torus, which is the group manifold $[U(1)]^6$, then A and B theories both reduce to N=8 supergravity in 4 dimensions.

The SO(6) frame symmetry of the extra 6 dimensions must contain an image of the stability group R of the space S/R. As explained by Salam and Strathdee[8], the vector of the frame SO(6) must be the same as the vector on S/R formed by the generators of S not in R. The essential point of coset-space dimensional reduction is that an R singlet is the only degree of freedom that can induce an S singlet in a harmonic expansion on S/R.

The catalogue of D=4 theories in Table I is derived by finding all 6-dimensional spaces S/R[14]. If R is trivial and S (or any nonabelian subgroup of S) has no image in the frame SO(6) symmetry, then the D=4 theory is N=8 supersymmetric. We neglect these cases here as they are well known. Slightly more interesting, but ignored in the Table I, are the many cases where the supersymmetry is completely destroyed and no spin 3/2 fields (or, indeed, any fermion) survive the reduction; that is, the $\underline{4}$ of SO(6) contains no R singlets. This is always the case if R is a symmetric subgroup of S or any subgroup of a symmetric subgroup of S. Some of these theories contain some number of scalars plus the graviton, where others couple the graviton, scalars and vectors. As an example, since U(1) is a symmetric subgroup of SU(2), the reduced theory on $[SU(2)/U(1)]XU(1)^4$ has no supersymmetry. [The branching rule of the $\underline{4}$ into U(1) representations is $\underline{4}$ = (1) + (1) + (−1) + (−1). Besides the graviton, there are 16 helicity 1 and 36 helicity 0 symmetric fields for both types A and B theories.] However, here we list only those cases where some supersymmetry is preserved. The problem is reduced to finding all spaces of 6 dimensions or less that are isomorphic to a coset space with a nonsymmetric stability group.

To see how the calculations are done, consider the coset space $G_2/SU(3)$. The embedding of SU(3) in G_2 is defined by the branching rule, $\underline{14} = \underline{8} + \underline{3} + \overline{\underline{3}}$, where the $\underline{14}$ is the adjoint representation of G_2. Since the $\underline{8}$ is the adjoint of SU(3), the vector transforms as $\underline{3} + \overline{\underline{3}}$ of SU(3). The other definition of the vector on $G_2/SU(3)$ is the $\underline{6}$ of the frame symmetry SO(6). Thus, the SU(3) must be embedded in SO(6) so that the $\underline{6} = \underline{3} + \overline{\underline{3}}$; this branching rule defines the embedding. The next step is to compute the SO(6) ⊃ R branching rules for the representations appearing in Table II, and then to count the R singlets for each helicity. The necessary SO(6) ⊃ SU(3) branching rules are: $\underline{4} = \underline{1} + \underline{3}$, $\underline{6} = \underline{3} + \overline{\underline{3}}$, $\underline{10} = \underline{1} + \underline{3} + \underline{6}$, $\underline{15} = \underline{1} + \underline{3} + \overline{\underline{3}} + \underline{8}$, $\underline{20} = \underline{3} + \overline{\underline{3}} + \overline{\underline{6}} + \underline{8}$, $\underline{20'} = \underline{6} + \overline{\underline{6}} + \underline{8}$. Thus the D=10 metric tensor (the $\underline{35}_v$) reduces to helicity 0, 2, and −2 fields, since the $\underline{6}$ and $\underline{20}'$ do not have SU(3) singlets. Proceeding, we find 1 symmetric helicity 2 field, 2 with helicity 3/2, 2 with helicity 1, 4 with helicity 1/2, and 6 with helicity 0. Once we remove the N=2 gravitational multiplet (with helicities 2, 3/2, 3/2, 1 plus conjugates) we have one helicity 1 state remaining, which is then the highest weight in a vector multiplet. Removing a vector multiplet and its conjugate, containing helicities, 1, 1/2, 1/2, 0, and 0, −1/2, −1/2, −1, we are then left with 2 helicity 1/2, 4 helicity 0 and 2 helicity −1/2 states, or two self-conjugate scalar multiplets.

In a similar fashion the rest of Table I is derived, and the remaining spaces for which the reduced theory has no supersymmetry can be checked. It is also amusing to reduce 11-dimensional simple supergravity by similar techniques, but with dim(S/R)=7. Besides the type A theories, D=4, N=1 theories can be found on spaces such as $SO(7)/G_2$ and SO(5)/SO(3).

Table I. D=4 Superfield Content of Dimensionally Reduced D=10, N=2 Supergravities. The type A theory can be obtained from D=11 simple supergravity; the type B theory is "chiral," and cannot be obtained from the 11-dimensional theory. The branching rules for the $\underline{4}$ of SO(6) defines the embedding for SO(6) ⊃ R. The vector on S/R is calculated from $\underline{6} = (\underline{4} \times \underline{4})_{antisymmetric}$.

S/R	Type A N=2, D=4 supermultiplets	Type B N=2, D=4 supermultiplets
G_2/SU(3) $\underline{4} = \underline{1} + \underline{3}$	1 gravitational 1 vector 2 scalars	1 gravitational 4 scalars
Sp(4)/SU(2)XU(1) 4 = 1(0) + 1(2) + 2(−1)	1 gravitational 2 vectors 2 scalars	1 gravitational 6 scalars
SU(3)/U(1)XU(1) $\underline{4}$ = (0,0) + (a,b) + (c,d) + (−a−c,−b−d)	1 gravitational 3 vectors 2 scalars	1 gravitational 8 scalars
SU(2)XSU(2)XSU(2)/SU(2) $\underline{4} = \underline{1} + \underline{3}$	1 gravitational 1 vector 4 scalars	1 gravitational 1 vector 4 scalars
	Type A N=4, D=4 supermultiplets	**Type B N=4, D=4 supermultiplets**
SU(2)XSU(2)XU(1)/U(1) $\underline{4}$ = (0) + (0) + (2) + (−2)	1 gravitational 4 N=4 vectors	1 gravitational 4 N=4 vectors

Table II. Branching Rules for SO(8) ⊃ SO(6) X U(1). The U(1) quantum number is normalized to the helicity in 4 dimensions.

$\underline{8}_v = \underline{1}(1) + \underline{1}(-1) + \underline{6}(0)$
$\underline{8}_s = \underline{4}(1/2) + \overline{\underline{4}}(-1/2)$
$\underline{8}_c = \overline{\underline{4}}(1/2) + \underline{4}(-1/2)$
$\underline{28} = \underline{1}(0) + \underline{6}(1) + \underline{6}(-1) + \underline{15}(0)$
$\underline{35}_v = \underline{1}(0) + \underline{20'}(0) + \underline{6}(1) + \underline{6}(-1) + \underline{1}(2) + \underline{1}(-2)$
$\underline{35}_c = \overline{\underline{10}}(1) + \underline{10}(-1) + \underline{15}(0)$
$\underline{56}_v = \underline{6}(0) + \underline{10}(0) + \overline{\underline{10}}(0) + \underline{15}(1) + \underline{15}(-1)$
$\underline{56}_s = \underline{4}(1/2) + \overline{\underline{4}}(-1/2) + \underline{20}(1/2) + \overline{\underline{20}}(-1/2) + \overline{\underline{4}}(3/2) + \underline{4}(-3/2)$
$\underline{56}_c = \overline{\underline{4}}(1/2) + \underline{4}(-1/2) + \overline{\underline{20}}(1/2) + \underline{20}(-1/2) + \underline{4}(3/2) + \overline{\underline{4}}(-3/2)$

* Talk delivered at the Eighth Johns Hopkins Workshop on Current Problems in Particle Theory, June 20-22, 1984 by R. Slansky.

**Work performed under the auspices of the U. S. Department of Energy under contract W-7405-ENG-48.

***Work performed in part by the U. S. Department of Energy under contract W-7405-ENG-36.

References

1. M. B. Green and J. H. Schwarz, Phys. Lett. 109B (1982) 444; 122B (1983) 143.
2. L. Alvarez-Gaume and E. Witten, Harvard Preprint HUTP-83/A039.
3. J. H. Schwarz, Phys. Reports 89 (1982) 223.
4. R. Slansky, Phys. Reports 79 (1981) 1.
5. R. Coquereaux and A. Jadczyk, Comm. Math. Phys. 90 (1983) 79; N. Manton, Santa Barbara Preprint NSF-ITP-83-04.
6. G. Chapline, Phys. Lett. 123B (1983) 401.
7. E. Cremmer and B. Julia, Nucl. Phys. B159 (1979) 141.
8. A. Salam and J. Strathdee, Ann. Phys. 141 (1982) 316.

9. M. J. Duff and C. N. Pope in "Supersymmetry and Supergravity '82," eds. S. Ferrara, J. G. Taylor and P. van Nieuwenhuizen (World Scientific Publishing, 1983).

10. M. Gunaydin and N. Warner, Caltech Preprint CALT-68-1077.

11. G. Chapline and G. W. Gibbons, Phys. Lett. 135B (1984) 43.

12. S. Coleman, "Classicial Lumps and Their Quantum Descendents," Appendix 4, 1975 Erice Proceedings.

13. J. Bagger and E. Witten, Nucl. Phys. B222 (1983) 1; G. Sierra and P. Townsend, Ecole Normale Preprint LPTENS 83/21 (May 1983); B. de Wit, P. Lauwers, R. Philippe, S. Q. Su, and A. Van Proeyen, NIKHEF preprint H/83-13 (July, 1983); M. Gunaydin, G. Sierra, and P. Townsend, Phys. Lett. 133B (1983) 72 and Ecole Normale preprint LPTENS 83/32 (1983).

14. G. Chapline and R. Slansky, Nucl. Phys. B209 (1982) 461. Note that Table 1 of this reference is restricted to spaces where the branching rule of the 4 of SO(6) in representations of R is not self conjugate. Thus, the list of spaces studied in the present paper is somewhat longer.

UCTP-104/84

Isometries As Probes of the Extra Dimensions*

Freydoon Mansouri and Louis Witten

Physics Department, University of Cincinnati
Cincinnati, Ohio 45221

ABSTRACT

We suggest that the consequences of dependence on the extra dimensions can best be studied in terms of isometries and their spontaneous symmetry breaking. We show explicitly how a dimensional reduction can be carried out in which isometries remove the traces of the extra dimensions from the reduced theory while ensuring that the solutions to the reduced theory are also solutions of the initial higher dimensional theory.

I. Introduction

The original ideas of Kaluza and Klein for unifying space-time and internal symmetries have been a source of inspiration for several generations of physicists. They also lead to geometries which, along with their generalizations, have been studied extensively in the mathematical literature. As a result, the myriads of contributions which have been made to this field vary both in emphasis and in physical content. The particular aspect which we wish to explore here is the development of quantitative methods to study the effects of extra dimensions. To put this objective in a proper perspective, it will be helpful to start by recalling some results from a selected number works representing various stages of this development. This selection is admittedly subjective and does not do justice to all the contributions made to this field; but it will hopefully provide some motivation for the work which will be described below.

The central point of the Kaluza-Klein approach is that the apparently distinct fundamental interactions in four dimensional space-time are remanents of a gravitational theory in a higher

*Presented by F. Mansouri

dimensional space-time. Thus, in this approach to unification, one starts with a manifold of more than four dimensions, in which for here-to-fore unknown reasons the dependence on all but three of the spacial dimensions disappear, or compactify, and certain internal symmetries appear instead. In so far as there is no physical or dynamical basis for this compactification, the formalism is flawed by the special properties asigned to the extra dimensions. It was, in particular, criticized by Einstein and Mayer[3] and by Einsitein and Bergmann[4], who did not regard this a genuine unification. This critisicism also applies in one way or another to much of the subsequent work on Kaluza-Klein Theories.

In the meantime, as the attention in theoretical physics was shifted away from unified theories, a substantial amount of technical progress was made in this field through the works of Cartan, Ehressmann, Kobayashi, etc. From the point of view of the present day interest in physics, we mention two of these. One is the Yang-Mills formulation of non-abalian gauge theories[5]. Properly speaking, the geometrical structure of such theories cannot be understood in four space-time dimensions: What we call a covariant derivative in a pure Yang-Mills theory with an n-parameter local gauge group is, in fact, a connection in a (4 + n)-dimensional manifold known as a principal fiber bundle. The other technical progress was the work of Lichnerowicz[6] in which he generalized the $5 \to 4$ dimensional reduction including the scalar field to $D \to D - 1$ and carried out successive reductions $D \to D - 1 \to D - 2$, etc.

The interest in dimensional reduction continued in the 1950's and 1960's at a reduced level. The next work we would like to mention is the work of Geroch[7]. He was interested in generating exact, source-free solutions of Einstein's equations, and, for any exact solution which has a killing vector, he was able to construct a one-paramenter family of solutions, each with a killing vector. The family is closed in the sense that the application of the transformation that he constructed to a complete circle of solutions results not in a new solution but in a rotation among the members of the family. Geroch later extended his results to the solutions having two commuting killing vectors. In this case he showed that by successive applications of his transformations one can find solutions which depend on a larger and larger number of parameters, eventually arriving at solutions which depend on arbitrary functions. Although these results were not intended to have any bearing on dimensional reduction in the sense used in Kaluza-Klein theories, they did amount to a dimensional reduction from $4 \to 3$ and $4 \to 2$, respectively. For example, in the case of one killing vector, the 4-dimensional vacuum Einstein equations become equivalent to a system of 3-dimensional Einstein equations coupled to a 3-dimensional vector and a 3-dimensional scalar. The important advantage of this approach is that the solutions of the reduced set of equations are also solutions of four dimensional Einstein's equations. As a result, the Einstein-Bergmann criticism does not apply here. It was this feature of the Geroch's work which inspired us to take a new look at dimensional reduction.

To proceed with this selective review, we want to mention that dimensional reduction has also been carried out in the framework reletivistic string models[8]. It is well known that the known bosonic and supersymmetric string models have critical dimensions D = 26 and D = 10, repsectively. What is not so well known is that, in the spirit of Kaluza-Klein theory, if one is willing to trade off some of the spacial dimensions with internal symmetries, then bosonic and supersymmetric string also exist in dimensions $D = 2 + 24/n$ and $D = 2 + 8/n$, respectively. Here the number n represents an SO(n-1) symmetry or the order of parastatistics in the paraquantization of the corresponding string model. In view of the recent progress in constructing a consistent supersymmetric string model[9], it would be of practical interest to see if its spectrum can have a 4-dimensional interpretation. For this to be true, one must set $n = 4$.

Turning again to works in the main stream of Kaluza-Klein approach, the recent revival of these theories originated in a couple of works which appeared in mid-seventies[10,11]. Further stimuli were provided by the extension of these works to coset spaces[12] and their application to N = 8 supergravity[13]. As of this writing, a considerable amount of information about these theories, their classical solutions, their quantum corrections, and their physical interpretation has become available. A number of hard conceptual problems remain, however. We would like to address some these in the sequel.

II. Our Aims and Ground Rules

Kaluza-Klein theories provide an attractive framework for the unification of internal and space-time symmetries. As is clear from the short review of the last section, some progress in this direction has been made, in particular in the context of supergravity theories. But to take this approach seriously, much remains to be done. This is because we do not yet have a systematic method for studying the effects of extra dimensions. Any such method must be able to provide satifactory answers to the following two questions: (1) Since there is no evidence for the existence of the extra dimensions at the shortest distances which can be probed, it must explain how this can be attributed to some intrinsic property of a higher dimensional theory. (2) It must provide a quantitative method for studying the consequences of the dependence on the extra dimensions.

In one of the approaches to dimensional reduction[14], it was pointed out that the complete absence of the dependence on the extra dimensions could be linked to the presence of exact isometries in the higher dimensional theory. Starting from this premise, one can then use the familiar spontaneous symmetry breaking techniques, which have been used successfully in particle physics, to relax, in a controllable manner, this nondependence. To see more explicitly how this can come about, we recall that in, e.g., electroweak or grand unified theories, the spin-1/2 and spin-1 sector is initially massless but has a large local symmetry as well as chiral symmetry. Then, the non-invariance of the vacuum under these symmetries leads to spontaneous symmetry breaking and the appearance of massive fermions and gauge bosons. In the same manner, if the physical vacuum is not invariant under a given initial isometry, then spontaneous symmetry breaking is expected to lead to the appearance of explicit dependence on the extra dimensions. Thus one has, at least in principle, a controllable method of studying the dependence on the extra dimensions.

In carrying out the above procedure, we must not lose sight of another feature of taking the extra dimensions seriously: The fundamental theory is the higher dimensional theory. Therefore, dimensional reduction must be carried out in such a way that the solutions to the resulting reduced theory will also be solutions to the original higher dimensional theory. It was to satisfy this condition that we adopted a generalization of Geroch's work for dimensional reduction. Thus, from our point of view the first objective of a dimensional reduction would consist in showing explicitly how isometries remove the traces of extra dimensions from the reduced theory while ensuring that the solutions to the reduced theory are also solutions of the initial higher dimensional theory.

III. The Details of Dimensional Reduction

To have a suitable formalism for dimensional reduction, we must generalize Geroch's method so that it would lend itself to reductions from any manifold, M, of dimension D, with or without torsion, to any manifold, S, of dimension $d < D$. The formalism must also be able to handle reductions associated with any isometry of M. To accomodate these requirements, we shall introduce vielbein fields and covariant derivatives in a way which is familiar from gauge theory approach to gravity and super-gravity[15]. This is equivalent to introducing a connection in the bundle of linear frames or the orthonormal frames with base manifold M.

Thus consider a D-dimensional manifold, M, with structure group SO(1,D-1). We specify a connection in M by the covariant derivative

$$\hat{D}_a = \partial_a + \hat{H}_a^{AB} \hat{X}_{AB} \tag{1}$$

$$a = o,..,D-1 \quad ; \quad A,B = 0,..,D-1$$

where $\hat{H}_a^{AB}$ are the gauge fields (connection coeffecients) and $\hat{X}_{AB}$ are (in general reducible) representations of the generators of the group SO(1,D-1). Here and in the sequel the "caret" on top of any quantity indicates that it corresponds to M, whereas the quantities without "caret" correspond to the reduced manifold S. For definiteness, we have assumed that M has one time and D-1 space dimensions. We also introduce a set of D-bein matrices $\hat{K}_a^A$ and their inverses $\hat{K}^b_B$, subject to the orthonormality conditions

$$\hat{K}^A_a \hat{K}^a_B = \delta^A_B \; ; \; \hat{K}^A_a \hat{K}^b_A = \delta^b_a \; . \tag{2}$$

One can then express the metric tensor on M in the form

$$\hat{g}_{ab} = \hat{K}^A_a \hat{K}^B_b \hat{\eta}_{AB} \tag{3}$$

where $\hat{\eta}_{AB}$ are the components of the Minkowskian metric with signature (+, -, .., -). It is also more convenient to express the covariant derivative (1) in a new basis given by

$$\hat{X}_A = \hat{K}^a_A \hat{D}_a \tag{4}$$

The set $\{\hat{X}_A\}$ satisfies commutation relations[15]

$$[\hat{X}_A, \hat{X}_B] = \hat{R}^{CD}_{AB} \hat{X}_{CD} - \hat{T}^{C}_{AB} \hat{X}_C \tag{5}$$

where $\hat{R}^{CD}_{AB}$ and $\hat{T}^{C}_{AB}$ are, respectively, the components of curvature and torsion in the basis $\hat{X}_A$.

Now suppose that the manifold M admits an n-parameter group of isometries. We assume that the corresponding n killing vectors $\overset{a}{\underline{\xi}}$ are everywhere space-like and satisfy the commutation relations

$$[\overset{i}{\underline{\xi}}, \overset{j}{\underline{\xi}}] = f_{ij}^{\ \ k} \overset{k}{\underline{\xi}} \; ; \quad i,j = 1,..,n \; . \tag{6}$$

With respect to the two bases defined above, the killing vectors have components $\{\overset{j}{\xi}{}^A\}$ and $\{\overset{j}{\xi}{}^a\}$ where

$$\overset{j}{\xi}{}^A = \hat{K}^A_a \overset{j}{\xi}{}^a \tag{7}$$

From (6) it follows that these components satisfy the relation

$$\overset{j}{\xi}{}^{A}\,\hat{X}_A\,\overset{k}{\xi}{}^{B} - \overset{k}{\xi}{}^{A}\,\hat{X}_A\,\overset{j}{\xi}{}^{B} = f_{jk}{}^{\ell}\,\overset{\ell}{\xi}{}^{B} + \overset{j}{\xi}{}^{A}\,\overset{k}{\xi}{}^{C}\,\hat{T}_{AC}{}^{B}\ . \tag{8}$$

By definition of an isometry, the metric tensor has a vanishing Lie derivative in the direction of the killing vectors:

$$\mathcal{L}_{\underset{\sim}{\overset{k}{\xi}}}\,\underline{g} = 0 \quad ; \quad k = 1,..,n\ . \tag{9}$$

This implies the generalized killing equation

$$\hat{X}_A\,\overset{k}{\xi}_B + \hat{X}_B\,\overset{k}{\xi}_A = \overset{k}{\xi}{}^{C}\,[\hat{\eta}_{AD}\,\hat{T}_{BC}{}^{D} + \hat{\eta}_{BD}\,\hat{T}_{AC}{}^{D}]\ . \tag{10}$$

To proceed with the reduction, it is convenient to restate the main objective of a dimensional reduction in a more down-to-earth form: it is to reexpress the field equations derived in M in the presence of isometries in terms of a set fields which could be interpreted as being defined on a lower dimensional manifold S. To see how this is carried out, let us divide M into orbits under the desired isometry. Two points p and q belong to the same orbit if there exists a curve from p to q the tangent to which at any point is a linear combination of the killing vectors. Thus the orbits are space-like hypersurfaces. We assume that they all belong to a single equivalence class. The elements of the manifold S are then taken to be the collection of all such orbits in M. It is clear from this construction that S is not a hypersurface in M but a quotient space of M by the action of isometry group. It becomes a hypersurface in the special case that the killing vectors are hypersurface orthogonal. We assume that the map from M to S exists and is smooth. Then, one can establish a one-to-one correspondence between tensor fields on S and tensor fields on M, which satisfy the following two conditions: (a) They are orthogonal to the killing vectors:

$$0 = \overset{j}{\xi}_A\,\hat{T}^{A..B}{}_{C..D} = .. = \overset{j}{\xi}_B\,\hat{T}^{A..B}{}_{C..D} = \overset{j}{\xi}{}^{C}\,T^{A..B}{}_{C..D} = .. = \overset{j}{\xi}{}^{D}\,T^{A..B}{}_{C..D}\ . \tag{11}$$

(b) They have vanishing Lie derivatives with respect to the killing vectors:

$$\mathcal{L}_{\underset{\sim}{\overset{j}{\xi}}}\,\underline{\hat{T}} = 0 \quad ; \quad j = 1,..,n\ . \tag{12}$$

This correspondence can be demonstrated formally in the manner sketched out by Geroch? We note, however, that the creation of the manifold S is purely for practical reasons and does not affect the process of dimensional reduction. In fact, one of the main advantages of our approach is that one need never abandon the original manifold, M, as the real space-time. Since our sence perceptions are used to interpreting the events as occuring in a d-(=4) dimensional space-time endowed with internal degrees of freedom, it is only convenient (and conforting) for the purposes of visualization, to refer to the manifold, S. From this point of view, the statement "fields on S," is a shorthand for "fields on M, which satisfy the conditions (a) and (b) above."

We begin the dimensional reduction by constructing a set of vielbeins on S. Consider the quantity

$$K^A_a = \hat{K}^A_a - \overset{j}{\xi}{}^A \, (\lambda^{-1})_{jk} \, \overset{k}{\xi}_a \quad . \tag{13}$$

It satisfies the conditions (a) and (b), so that it may be identified as the vielbein on S. In this expression $\lambda^{-1}{}_{jk}$ is the inverse of the matrix

$$\lambda_{jk} = \overset{j}{\xi}{}^A \, \overset{k}{\xi}_A \quad . \tag{14}$$

From the vielbeins we can construct the components of the kronecker delta and metric tensors on S:

$$h^B_A = K^B_b \, K^b_A = \delta^B_A - \overset{j}{\xi}{}^B \, (\lambda^{-1})_{jk} \, \overset{k}{\xi}_A \tag{15}$$

$$h^b_a = K^A_a \, K^b_A = \delta^b_a - \overset{j}{\xi}{}^b \, (\lambda^{-1})_{jk} \, \overset{k}{\xi}_a \tag{16}$$

$$h_{ab} = \eta_{AB} \, K^A_a \, K^B_b = \hat{g}_{ab} - \overset{j}{\xi}_a \, (\lambda^{-1})_{jk} \, \overset{k}{\xi}_b \tag{17}$$

$$h_{AB} = h_{ab} \, K^a_A \, K^b_B = \hat{\eta}_{AB} - \overset{j}{\xi}_A \, (\lambda^{-1})_{jk} \, \overset{k}{\xi}_B \tag{18}$$

It is instructive to solve equation (17) for $\hat{g}_{ab}$:

$$\hat{g}_{ab} = h_{ab} + \xi^{j}_{a} (\lambda^{-1})_{jk} \xi^{k}_{b} \ . \tag{19}$$

Comparing this with the expression for the components of the metric tensor for a fiber bundle given in standard texts on differential geometry, we have an explicit proof that the manifold M in presence of an isometry has indeed the structure of a fiber bundle. Therefore, locally, and only locally, these manifolds have the structure S x F where the fiber (orbit) F is a compact manifold.

A connection on S can be induced from a connection on M. Thus, given a tensor on M satisfying the conditions (a) and (b) above, the action of the covariant derivative X_A on this tensor can be deduced from the action of $\hat{X}_A$ by using the projection operator (15):

$$X_A T^{B..C}_{D..E} = h^{L}_{A} h^{M}_{D} ..h^{N}_{E} h^{B}_{I} ..h^{C}_{J} \hat{X}_L T^{I..J}_{M..N} \ . \tag{20}$$

Since, the connections on M and S are both nonsymmetric, in general, and since the nonsymmetric part of the connection is determined by the torsion tensor, we must demand that this tensor is compatible with the structure of (20). To ensure this, we require that the torsion tensor of M be a tensor on S:

$$\xi^{A}_{k} \hat{T}^{C}_{AB} = \xi^{B}_{k} \hat{T}^{C}_{AB} = 0$$
$$\mathcal{L}_{\underline{\xi}_k} \underline{T} = 0 \ . \tag{21}$$

It can then be verified that $\{X_A\}$ satisfy all the requirements for a covariant derivative on S and that they satisfy the commutation relations

$$[X_A , X_B] = R^{CD}_{AB} X_{CD} - T^{C}_{AB} X_C \tag{22}$$

where R_{AB}^{CD} and T_{AB}^{C} are, respectively, the components of curvature and torsion tensors of S, and X_{CD} are the generators of SO(1, d-1).

IV. Spectra and Equations of Motion

In harmony with our view of taking the extra dimensions seriously, we start by postulating an action in the D-dimensional manifold, M. For definiteness, suppose we take the action to be the Einstein-Hilbert-Cartan action. The variation of this action leads to the field equations

$$\hat{R}^{A}_{B} = 0 \quad . \tag{23}$$

A dimensionally reduced theory is then a solution of these equations for a given group of isometries. Moreover, such solutions must be expressible in terms of a set of fields on S. The components, h_{ab}, of the metric tensor on S have, by construction, the required properties to belong to this set. They are not enough, however, and we must look for additional fields which satisfy our conditions (a) and (b) to complete the set. From the structure of $\hat{g}_{ab}$ as given by equation (19), it is clear that the additional fields must some how be related to the killing vectors. For abelian isometries, the required quantities can, in fact, be constructed from the killing vectors. For example, the fields λ_{jk} given by (14) have vanishing Lie derivatives and are scalars on S. But for non-abelian isometries, λ_{jk} are no longer scalars on S:

$$\mathcal{L}_{\overset{m}{\underline{\xi}}} \lambda_{jk} \neq 0 \quad \text{when} \quad f_{ij}^{\;\;k} \neq 0 \quad . \tag{24}$$

We must, therefore, look for vector fields, $\overset{k}{\underline{\Psi}}$, which satisfy the same commutation relations as $\overset{k}{\underline{\xi}}$, i.e.,

$$[\overset{j}{\underline{\Psi}}, \overset{k}{\underline{\Psi}}] = f_{jk}^{\;\;m} \overset{m}{\underline{\Psi}} \quad , \tag{25}$$

but are left invariant by the isometry group:

$$[\overset{j}{\underline{\Psi}}, \overset{k}{\underline{\xi}}] = 0 \; ; \; j,k = 1,..,n \quad . \tag{26}$$

Since the killing vectors are linearly independent, the vectors $\overset{k}{\underline{\Psi}}$, when they exist, can be written as a linear combination of the killing vectors with <u>non-constant</u> coefficients

$$\overset{j}{\underline{\Psi}}(z) = \alpha_{jk}(z)\,\overset{k}{\underline{\xi}}(z) \tag{27}$$

the Z^A, A = 0, , D - 1, are the coordinates of M. From equations (25) - (27), it follows that

$$\overset{m}{\underline{\xi}}\,\alpha_{ij} = f^{\,j}_{km}\,\alpha_{ik} \quad ; \quad m,i,j = 1,..,n \ . \tag{28}$$

In a coordinate basis, this gives a system of differential equations which can be solved for the coefficients α_{ij}. The existence of such solutions for groups of interest can be established by general arguments. Although the explicit forms of α_{ij} do not enter actual computations, it is instructive to see how equations (28) can be solved for a simple example. Thus consider the 2-parameter non-abelian isometry group for which the killing vectors satisfy the commutation relations

$$[\,\overset{1}{\underline{\xi}}\,,\,\overset{2}{\underline{\xi}}\,] = \overset{1}{\underline{\xi}} \quad ; \quad \overset{1}{\underline{\xi}}\,,\,\overset{2}{\underline{\xi}} \text{ everywhere space-like} \ . \tag{29}$$

In a coordinate basis, one can take

$$\overset{1}{\underline{\xi}} = \frac{\partial}{\partial x} \quad ; \quad \overset{2}{\underline{\xi}} = x\frac{\partial}{\partial x} + \frac{\partial}{\partial y} \ . \tag{30}$$

Then equations (28) reduce to

$$\begin{aligned} \frac{\partial}{\partial x}\,\alpha_{ij} &= f^{\,j}_{k1}\,\alpha_{ik} \\ \left(x\frac{\partial}{\partial x} + \frac{\partial}{\partial y} \right)\alpha_{ij} &= f^{\,j}_{k2}\,\alpha_{ik} \ . \end{aligned} \tag{31}$$

A general solution of these equations is

$$(\alpha_{ij}) = \begin{bmatrix} e^y & 0 \\ -\beta x & \beta \end{bmatrix} \tag{32}$$

where β is an arbitrary constant. From (27), we find

$$\begin{aligned} \overset{1}{\underline{\Psi}} &= e^y \frac{\partial}{\partial x} \\ \overset{2}{\underline{\Psi}} &= \beta \frac{\partial}{\partial y} . \end{aligned} \tag{33}$$

Clearly, $[\overset{j}{\underline{\Psi}}, \overset{k}{\underline{\xi}}] = 0$, and for $\beta = -1$, $[\overset{1}{\underline{\Psi}}, \overset{2}{\underline{\Psi}}] = \overset{1}{\underline{\Psi}}$.

As pointed out above, in practice the knowledge of the explicit forms of α_{ij} are most often unnecessary. Instead, it is more convenient to make use of a number of identities which are satisfied by them. Some of these are

$$\begin{aligned} \alpha_{jm}\, \alpha_{ks}\, f_{sm}^{\ r} &= f_{jk}^{\ m}\, \alpha_{mr} \\ \alpha_{jm}\, f_{km}^{\ r}\, \alpha^{-1}_{rs} &= \alpha^{-1}_{kr}\, f_{jr}^{\ s} \\ \alpha_{jm}\, f_{mk}^{\ k} &= -f_{jk}^{\ k} \end{aligned} \tag{34}$$

$$\hat{X}_A\, \alpha_{jk} = \overset{r}{\xi}_A\, \lambda^{-1}_{rs}\, f_{ms}^{\ k}\, \alpha_{jm} . \tag{35}$$

With vector fields $\overset{k}{\underline{\Psi}}$ at our disposal, we can now proceed to complete the list the fields on S which determine the reduced theory. It is easy to verify that the scalar fields

$$\mu_{jk} = \overset{j}{\underline{\Psi}} \cdot \overset{k}{\underline{\Psi}} = \overset{j}{\Psi}{}^A\, \overset{k}{\Psi}_A \tag{36}$$

have vanishing Lie derivatives with respect to the killing vectors $\overset{m}{\xi}$. They are, therefore, scalar fields on S. It is also easy to show that

$$\overset{j}{\xi}{}^B\, (\lambda^{-1})_{jk}\, \overset{k}{\xi}_A = \overset{j}{\Psi}{}^B\, (\mu^{-1})_{jk}\, \overset{k}{\Psi}_A . \tag{37}$$

Therefore, in equations (15) - (19) the dependence on $\xi^{j}{}_{A}$'s and $(\lambda^{-1})jk$'s can be expressed in terms of $\psi^{j}{}_{A}$'s and $(\mu^{-1})jk$'s. For example,

$$\hat{g}_{ab} = h_{ab} + \psi^{j}{}_{a} (\mu^{-1})_{jk} \psi^{k}{}_{b} \quad . \tag{38}$$

Finally, to complete the list of our fields on S, we define n antisymmetric fields ((d-2) - forms) $\omega_{j}^{A..B}$ according to

$$\omega_{j}^{A..B} = [n!\ (d-2)!]^{-1}\ \varepsilon_{\ell\cdot\cdot m}\ \varepsilon^{A..BC..DEF}\ \psi^{\ell}{}_{C} \cdot\cdot \psi^{m}{}_{D} \hat{X}_{E} \psi^{j}{}_{F} \tag{39}$$

where ε's are the appropriate Levi- Civita tensors. It is straight-forward to show that

$$\xi^{k}{}_{A}\ \omega_{j}^{A..B} = \cdot\cdot = \xi^{k}{}_{B}\ \omega_{j}^{A..B} = 0 \tag{40}$$

$$\mathcal{L}_{\underline{\xi}^{k}}\ \omega_{j}^{A..B} = 0 \tag{41}$$

so, $\omega_{j}^{A..B}$ are tensors on S.

A central result of this work is that given the fields h_{ab}, μ_{ij} and $\omega_{j}^{A..B}$, it is possible to express the gravitational field equations (23) in the presence of an isometry of M in tems of a set of equations for these fields, to which we now turn. For the divergences and the curls of the (d-2)-forms, we get, respectively,

$$\begin{aligned} X_{A}\omega_{j}^{AA'..BB'} &= \mu^{-1}_{\ell m}\ [\omega_{m}^{A..B}\ X_{A}\mu_{\ell j} + \tfrac{1}{2}\omega_{j}^{A..B} X_{A}\mu_{\ell m}] \\ &+ \omega_{j}^{A..B}\ T_{FA}{}^{F} + \tfrac{1}{2}(-1)^{d}\ \omega_{j}^{AE[A..B}\ T^{B]}{}_{AE} \end{aligned} \tag{42}$$

$$\begin{aligned} X^{[H}\omega_{j}^{A..BB]} &= \omega_{j}^{[A..B}\ \hat{\eta}^{H]F}\ T^{E}{}_{EF} + (-1)^{d}(d-2)\,\omega_{j}^{M[A..B}\ T_{MF}{}^{B}\ \hat{\eta}^{H]F} \\ &+ [(n!)(d-1)!]^{-1}\ \mu^{-1}_{eb}\ \varepsilon_{\ell p..qh}\ \varepsilon^{A..BCC'..DDEH}\ \psi^{p}{}_{C'} \cdot\cdot \psi^{q}{}_{D'}\ x \\ &x\,\{nX_{E}\mu_{je}[f_{km}{}^{m}\psi^{\ell}{}_{C}\ \psi^{b}{}_{D} - \tfrac{1}{2}(n-1)\psi^{m}{}_{C}\ \psi^{b}{}_{D} f_{k\ell}{}^{m} + \psi^{\ell}{}_{C}\psi^{d}{}_{D} f_{bk}{}^{d}]\} + \psi^{\ell}{}_{C}\psi^{k}{}_{D} f_{bj}{}^{d}\ X_{E}\ \mu_{ed}\} . \end{aligned} \tag{43}$$

In these expressions, the brackets [A..B] mean antisymmetrization with respect to the correspoonding indices. For compact Lie groups, the structure constants $f_{ij}^{\ \ k}$ can be taken to be anti-symmetric in all indices, so that $f_{km}^{\ \ \ m} = 0$. But for groups of the type discussed in the example above, $f_{km}^{\ \ \ m} \neq 0$. For the scalar fields we get

$$
\begin{aligned}
X^A X_A\, \mu_{ab} &= \mu^{-1}_{cd}(X^A \mu_{ac})(X_A \mu_{bd}) - \tfrac{1}{2}\mu^{-1}_{cd}(X^A \mu_{cd})(X_A \mu_{ab}) \\
&+ (-1)^{d+n-1}[(d-2)!]\mu^{-1}\omega_{aI..J}\omega_b^{I..J} + f_{dk}^{\ \ k}\mu^{-1}_{cd}(f_{cb}^{\ \ m}\mu_{am} + f_{ca}^{\ \ m}\mu_{bm}) \\
&+ \mu^{-1}_{cd}[2\mu_{ij} f_{ad}^{\ \ i} f_{bc}^{\ \ j} + 3\mu_{ia} f_{dk}^{\ \ i} f_{cb}^{\ \ k} + 3\mu_{ib} f_{dk}^{\ \ i} f_{ca}^{\ \ k}] - 4 f_{ad}^{\ \ m} f_{bm}^{\ \ d} \qquad (44) \\
&+ \mu^{gh}_{pq}(\mu_{ia} f_{gh}^{\ \ i} + 2\mu_{ig} f_{ah}^{\ \ i})(f_{qp}^{\ \ m}\mu_{mb} + 2\mu_{mq} f_{bp}^{\ \ m}) \\
&+ \frac{1}{8}\mu^{dh}_{pq}\mu^{eg}_{rs}\mu_{de}\mu_{hg}(f_{qp}^{\ \ i}\mu_{ia} + 2 f_{ap}^{\ \ i}\mu_{iq})(f_{sr}^{\ \ m}\mu_{mb} + 2 f_{br}^{\ \ m}\mu_{ms})
\end{aligned}
$$

In this expression, $\mu = \det \mu_{ij}$ and

$$\mu^{\ell k}_{pq} = [\mu(n-2)!]^{-1}\varepsilon^{h..j\ell k}\varepsilon_{r..spq}\,\mu_{hr}..\mu_{js}\,. \qquad (45)$$

Finally, the gravitational field equations for the metric h_{ab} take the form

$$
\begin{aligned}
R^M_B &= \frac{1}{4}(X^M \mu_{ab})(X_B \overset{-1}{\mu}_{ab}) + \tfrac{1}{2}\mu^{-1}_{ab} X^M X_B \mu_{ab} - \tfrac{1}{2}\mu^{-1}_{ab}\eta^{MA} T_{BA}^{\ \ E} X_E \mu_{ab} \\
&+ \tfrac{1}{2}(-1)^{d+n-1}(d-2)!\,\mu^{-1}\mu^{-1}_{ab}[h^M_B\, \omega_{aA..E}\omega_b^{A..E} - (d-2)\omega_{aB..A}\omega_b^{M..A}] \qquad (46) \\
&+ (-1)^{d+n}[2\mu(n!)]^{-1}\mu^{-1}_{jk} f_{jm}^{\ \ m}\varepsilon_{p..q}\varepsilon_{K..LG..HNB}\,\psi^{pG}..\psi^{qH}\omega_k^{K..L}\eta^{MN}\,.
\end{aligned}
$$

It cannot be overemphasized that equations (42), (43), (44) and (46) are consequences of the variation of an action not in d- but in D-dimensions. As a result, their solutions are also solutions of the initial higher dimensional theory. For completeness, we also give the expression for the scalar curvature R of space M:

$$
\begin{aligned}
\hat{R} &= R - \tfrac{1}{2}\mu^{-1}_{ab} X^A X_A \mu_{ab} - \tfrac{1}{4}(X^A \mu_{ab})(X_A \mu^{-1}_{ab}) \\
&\quad + (-1)^{d+n}[(d-2)!]\mu^{-1}\mu^{-1}_{ab}\,\omega_{aA..B}\,\omega_b^{A..B}\,. \qquad (47)
\end{aligned}
$$

In the conventional approach to Kaluza-Klein theories, a Lagrangian density based on $\hat{R}$ is taken as a starting point. In contrast to the solutions of our equations, the solutions of the field equations which result from such theories are not, in general, also solutions of (23). They will, therefore, be subject to a generic Einstein-Bergmann criticism.

V. Discussion

In the previous sections, we have seen how it is possible to have a D-dimensional theory which, in the presence of isometries, looks like a d-dimensional theory with $d < D$. Just as the masslessness of certain quanta in a field theory can be attributed to the presence of symmetries, the non-appearance of the (D - d)—dimensions can be attributed to the presence of an isometry of the original theory. Then if the extra dimensions really exist, it is reasonable to expect that the consequences of the dependence on these dimensions can be studied by breaking the isometry which is responsible for their non-appearance. We would like to give a brief description of how this can come about by stating what we mean by symmetry breaking in the context of dimensional reduction.

It will be recalled that in conventional gauge theories, spontaneous symmetry breaking takes place at a stage when there is a clear distinction between space-time and internal space, so that it results in the reduction of the symmetry associated with internal degrees of freedom while keeping the space-time symmetry intact. Clearly, a straight-forward application of this technique to any dimensionally reduced theory in which the space-time and internal space have already been separated will not yield any information about the extra dimensions. To learn about these dimensions, one must go back to the original D-dimensional theory and consider all possible isometries of the manifold M. With one time and (D-1) space dimensions, the maximum number of the one parameter groups of isometries of M is limited by its structure group GL (DR). Since isometries are metric preserving transformations, this implies that the largest group of isometries of the manifold M is SO(1, D-1). The largest subgroup of this group which can be associated with an internal symmetry is determined by the choice of space-time. For example, in the fiber bundle solutions discussed above, we wish to identify the d-dimensional base manifold, S, as the dimensionally reduced space-time. If we wish to retain the structure group SO(1, d-1) as the symmetry of such a space-time, then of the isometries of the manifold M, the part which can be devoted to internal symmetries is contained in the quotient SO(1, D-1)/SO(1, d-1). To be a local isometry, this quotient must itself be a group or contain a group. In this way, one arrives at SO(D-d) as the largest available isometry group for internal symmetry.

As mentioned above, if we start with an isometry group of the form SO(1, d-1) $\otimes$ SO(D-d) and proceed to subject SO(D-d) to spontaneous symmetry breaking, we will end up getting the conventional results which are familiar from, say, the grand unified models. In such an approach there is no room for the effects related to the extra dimensions. To see how the extra dimensions lead to physical consequences, we must allow for the possibility that the dimension, d, of the reduced manifold varies in the process of spontaneous symmetry breaking. Then, for fixed D, as d increases, the internal symmetry group, SO(D-d), is broken down to a smaller group and, simultaneously, the space-time symmetry group, SO(1, d-1) increases. This is as it should be because by taking one or more of the extra dimensions as real, we are considering a real world with a larger number of spacial dimensions and hence with a larger structure group. For example, if $d \to d + 1$, $SO(1, d-1) \otimes SO(D-d) \to SO(1, d) \otimes SO(D-d-1)$. Thus, in this context, it is more appropriate to speak of "Phase transition" or "symmetry transmutation", than of plane spontaneous symmetry breaking. These possibilities are currently being studied.

We would like to thank Peter Suranyi for discussions. This work was supported in part by Department of Energy under Contract No. DE-AS-2-76ER02978.

References

1. Th. Kaluza, Sitzungsber. Preuss Acad. Wiss. Berlin, math-Phys. K1, 966 (1921).

2. O. Klein, Z. Phys. 37, 895 (1926).

3. A. Einstein and W. Mayer, Preuss. Akad. P. 541 (1931) and P. 130 (1932).

4. A. Einstein and P. Bergmann, Ann. of Math 39, 683 (1938).

5. C.N. Yang and R.L. Mills, Phys. Rev. 96, 191 (1954).

6. A. Lichnerowicz, Theories Relativistes de la Gravitation et de l' Electromagnetisme (Masson, Paris, 1955).

7. R. Geroch, J. Math Phys. 12, 918 (1971); 13, 394 (1972); F.P. Esposito and L. Witten, Proceedings of Second Latin American Symposium on General Relativity and Gravitation, Carcas, 1976, ed. C. Aragone, Simon Bolivar University.

8. F. Ardalan and F. Mansouri, Phys. Rev. D9, 3341 (1974).

9. M. Green and J. Schwarz, Nucl. Phys. B181, 502 (1981).

10. L.N. Chang, K. Macrae, and F. Mansouri, Phys. Lett. B57, 59 (1975); Phys. Rev. D13, 235 (1976).

11. Y.M. Cho and P.G.O. Freund, Phys. Rev. D12, 1711 (1975).

12. G. Domokos and S. Kovesi-Domokos, Nuovo Cimento

44A, 318 (1978).

13. E. Cremmer and B. Julia, Phys. Lett. B80, 48 (1978).

14. F. Mansouri and L. Witten, Phys. Lett B127, 341 (1983), B140, 317 (1984); Jour. Math Phys. 25, 1991 (1984).

15. F. Mansouri, Proceedings of VIII International Colloquium on Group Theoretical Methods in Physics, ed. W. Beiglbuck, A. Bohm, and E. Takasugi (Springer-Verlag, Berlin, 1978); F. Mansouri and C. Schaer, Phys. Lett. B101, 51 (1981).

GAUGE THEORY IN CURVED SPACE

Yutaka Hosotani

Department of Physics, University of Pennsylvania
Philadelphia, Pennsylvania 19104

ABSTRACT

Non-Abelian gauge theory in curved space is analysed from two standpoints. In the first half various exact classical solutions are presented which solve matter field equations and the Einstein equations simultaneously. Their implications to the very early universe are discussed. In the second half the problem of dynamical mass generation by finite curvature is investigated to show the importance of quantum effects to phase transitions in curved space.

1. Introduction

Our present understanding of Nature consists of two backbones, non-Abelian gauge theory of electroweak and strong interactions and Einstein's general relativity. Both of them are characterized by highly non-linear equations, having been subject to extensive investigation. In this talk I combine these two to analyse gauge theory in curved space.

It is well known that the evolution of the universe is well described by the Friedmann-Robertson-Walker model, in which spatial curvature becomes larger and larger as one goes back to the earlier time. In particular, if the universe underwent an exponential expansion in the very early time as has been strongly suggested in the inflationary universe scenario,[1] spatial curvature of the universe in the preinflation period can be quite large, in no contradiction to the observed extreme flatness of the present universe. It is in this regime that strong gravity intertwines high nonlinearity of the non-Abelian gauge theory to lead to many interesting phenomena.

In this talk I would like to focus on two particular aspects. In the first half various exact classical solutions are examined which solve matter field equations and the Einstein equations simultaneously. It will be seen that many simple, interacting theories admit not only a flat Minkowski spacetime solution, but also non-trivial curved-spacetime solutions. Their implications to the very early universe are briefly discussed. In the second half I investigate the quantum aspect of non-Abelian gauge theory in background gravitational fields, especially focusing on the problem of dynamical mass generation by finite curvature.

2. Exact Solutions. I. Conformally invariant theories

2-A Conformal invariance

In general it is a difficult task to find an exact classical solution which solves both matter field equations and the Einstein equations simultaneously. The problem is considerably simplified in conformally invariant theories such as a scalar field theory

$$I_S = \int d^4x \sqrt{-g}\, \{\tfrac{1}{2} g^{\mu\nu} \partial_\mu\phi\partial_\nu\phi - \tfrac{1}{12} R\phi^2 - \tfrac{\lambda}{4!}\phi^4\} \tag{1}$$

and a Yang-Mills gauge field theory

$$I_{YM} = \int d^4x \sqrt{-g}\, (-\tfrac{1}{4})\, F^a_{\mu\nu} F^{a\mu\nu} . \tag{2}$$

Both I_S and I_{YM} are invariant (up to a surface term in I_S) under conformal (Weyl) transformations

$$\begin{aligned} g_{\mu\nu} &\to \tilde{g}_{\mu\nu} = \Omega(x)^{-2}\, g_{\mu\nu} \quad , \\ \phi &\to \tilde{\phi} = \Omega(x)\phi \quad , \\ A_\mu &\to \tilde{A}_\mu = A_\mu \quad . \end{aligned} \tag{3}$$

With applications to the early universe in mind we seek for a self-consistent solution in which a metric takes the Robertson-Walker form:

$$ds^2 = a(\eta)^2 \{d\eta^2 - \frac{dr^2}{1-kr^2} - r^2 d\Omega^2\} \quad , \ k = 0, \pm 1. \tag{4}$$

A solution is found in three steps. (i) First we solve matter equations in a static metric ($a(\eta)=1$ in (4)) to find $\tilde{\phi}$ and $\tilde{A}$. (ii) Second, we make a conformal transformation to return to the original metric (4), thereby $\tilde{\phi}$ and $\tilde{A}_\mu$ being transformed into $\phi = \tilde{\phi}/a(\eta)$ and $A_\mu = \tilde{A}_\mu$, respectively. Conformal invariance guarantees that ϕ and A_μ solve matter equations in the metric (4). (iii) Finally, $a(\eta)$ is determined by solving the Einstein equations.

2-B Scalar field solution

A solution has been found in an open universe case (k=-1) by Frolov, Grib, and Mostepanenko.[2] A scalar field equation is

$$\phi_{;\mu}{}^{\mu} + \frac{1}{6} R\phi + \frac{\lambda}{3!}\phi^3 = 0. \tag{5}$$

In a static metric $\tilde{g}_{\mu\nu}$

$$ds^2 = d\eta^2 - \frac{dr^2}{1+r^2} - r^2 d\Omega^2$$

the scalar curvature $\tilde{R}$ is negative ($\tilde{R}=-6$). Since $\frac{1}{6}R$ in Eq. (5) effectively acts as $(\text{mass})^2$, we find a non-trivial solution $\tilde{\phi} = \pm(3!/\lambda)^{\frac{1}{2}}$. In the original metric

$$\phi = \pm\sqrt{\frac{3!}{\lambda}}\,\frac{1}{a(\eta)} \quad . \tag{6}$$

The Einstein equations resolve, with (6) substituted into their r.h.s., into two equations for $a(\eta)$:

$$a'' = a,$$
$$a^2 - a'^2 = \frac{4\pi}{\lambda} G \equiv a_o^{\,2} \quad , \tag{7}$$

where G is the gravitational constant. The solution is

$$a(\eta) = a_o \cosh(\eta - \eta_o) \quad ,$$

or, in terms of the proper synchronous time t ($dt = a d\eta$),

$$a(t) = \{a_o^{\,2} + (t - t_o)^2\}^{\frac{1}{2}} \quad , \tag{8}$$

where $t_o = a_o \sinh \eta_o$.

A few comments are in order. In the solution the energy-momentum tensor is traceless ($T^\mu{}_\mu = 0$), but $T_o{}^o$ is negative.

$$T_o{}^o = -\frac{1}{4}\frac{3!}{\lambda}\frac{1}{a(t)^4} < 0 \quad . \tag{9}$$

This configuration is energetically more favored than a flat Minkowski spacetime solution ($g_{\mu\nu} = \eta_{\mu\nu}$, $\phi=0$). However, energy density is not a good criterion for stability in curved space. Further analysis is necessary to see which one is more stable than the other. If a scalar field is coupled to gauge fields as in the standard unified theory of electroweak and strong interactions, non-vanishing ϕ in (6) means that gauge symmetry is always broken in the solution.

2-C Yang-Mills field solution

Recently a non-trivial solution in the system I_{YM}, (2), has been found in a closed universe case (k=+1) by the author[3]. Topology of spacetime in a closed universe is $R^1 \times S^3$. As is well known, there is one-to-one correspondence between points in S^3 and group elements in SU(2):

$$S^3: \quad y_1^{\,2} + y_2^{\,2} + y_3^{\,2} + y_4^{\,2} = r^2$$
$$\leftrightarrow h = (y_4 + i\vec{y}\cdot\vec{\tau})/r \in SU(2) \quad , \tag{10}$$

where τ^j's are Pauli matrices. Upon noticing this natural mapping, one can easily find a non-trivial exact solution.

Since S^3 is a group manifold, it is most convenient to use Maurer-Cartan one forms as a basis of tetrads in S^3. Left-invariant Maurer-Cartan one forms σ^j are defined by

$$h^{-1}dh = i\vec{\sigma}\,\vec{\tau} \quad , h \in SU(2) \quad . \tag{11}$$

$$d\sigma^i = \epsilon^{ijk}\sigma^j \wedge \sigma^k \quad . \tag{12}$$

In terms of σ^j's a line element in a closed Robertson-Walker space is given by

$$ds^2 = a(\eta)^2 \{d\eta^2 - d\ell^2(S^3)\} \quad ,$$
$$d\ell^2(S^3) = \sigma^j{}_m \sigma_{jn} \, dx^m dx^n \quad , \tag{13}$$

where $\sigma^j = \sigma^j{}_m \, dx^m$.

The Yang-Mills equations in SU(2) gauge theory are

$$F^{\mu\nu}{}_{;\nu} - ig[A_\nu, F^{\mu\nu}] = 0 \quad . \tag{14}$$

In terms of differential forms Eq. (14) can be expressed as

$$d(*F) - ig(A\wedge *F \mp *F\wedge A) = 0 \quad . \tag{15}$$

where $A = A_\mu dx^\mu$ and $F = dA - igA\wedge A$. The sign in front of the last term is negative (positive) in even (odd) dimensions.

A solution to Eq. (15) is found in a similar fashion to that in the previous case. In a static metric we make an Ansatz

$$A = -\frac{f}{2g} \vec{\sigma} \vec{\tau} \, , \tag{16}$$

which means that everything is time-independent and $A_o = 0$ so that the Yang-Mills equations reduce to those in three dimensions. Insertion of (16) into Eq. (15) yields

$$-\frac{1}{2g} f(f-1)(f-2)\varepsilon^{ijk}\sigma^i\wedge\sigma^j \cdot \tau^k = 0 \quad , \tag{17}$$

which is solved by f = 0, 1, and 2. f = 0 and 2 correspond to pure gauge configurations ($F \propto f(f-2)$), whereas f=1 gives a non-trivial solution.

Had one used right-invariant Maurer-Cartan one forms $\bar{\sigma}^j$ ($dh\cdot h^{-1} \equiv -i\bar{\sigma}^j\tau^j$) instead of (11), one would have obtained a solution $A = -(1/2g)\bar{\sigma}^j\tau^j$, which is gauge equivalent to $A = -(1/2g)\sigma^j\tau^j$: $A = hAh^{-1} - (i/g)dh\cdot h^{-1}$.

The solution in the original metric is obtained by conformal transformation. In a tetrad system ($e^o = ad\eta$, $e^j = a\sigma^j$), non-vanishing components of A and F are

$$A^p_j = e_j{}^\mu A^p_\mu = -\frac{1}{ga(\eta)} \delta_{pj},$$
$$F^p_{jk} = -\frac{1}{ga(\eta)^2} \varepsilon^{pjk} \quad , \tag{18}$$

the index p referring to the internal SU(2) space.

$a(\eta)$ is determined by the Einstein equations, which resolve into two equations:

$$a'' + a = 0$$
$$a'^2 + a^2 = \frac{4\pi G}{g^2} \equiv a_o{}^2 \quad . \tag{19}$$

The solution is

$$a = a_o \sin\eta = t^{\frac{1}{2}}(2a_o - t)^{\frac{1}{2}}, \tag{20}$$

where $dt = a d\eta$.

The solution is purely magnetic and breaks SU(2) gauge symmetry completely. It corresponds to a closed universe with the energy-momentum tensor dominated by large magnetic fields. The lifetime of the universe is $2a_o$ (too short!).

The stability of the solution against small fluctuations of gauge fields can be completely analysed. It turns out that there is one, and only one, unstable mode

$$\delta A^p_j = \text{const.} \times \frac{1}{ga(\eta)} \quad e^{\sqrt{2}\,\eta}\,\delta_{pj} \quad . \tag{21}$$

Note that $\eta = \cos^{-1}(1-t/a_o)$ and $0 \le \eta \le \pi$. The unstable mode grows at most by a factor $\exp(\sqrt{2}\,\pi) \sim 80$.

3. Exact solutions II. Conformally non-invariant theories

If a theory is not conformally invariant, its energy-momentum tensor is not traceless in general so that the theory may admit a de Sitter space solution. We consider a scalar field theory described by

$$I = \int d^4x \sqrt{-g}\, \{\tfrac{1}{2} g^{\mu\nu}\partial_\mu\phi\partial_\nu\phi - \tfrac{1}{2}\xi R\phi^2 - V[\phi] + \frac{1}{16\pi G} R\} \tag{22}$$

We impose a requirement on $V[\phi]$ such that a flat Minkowski space be a solution to (22), i.e., $V[\phi]$ must vanish at least at one of its minima.

3-A. Trivial solutions ($\xi = 0$)

If $V[\phi]$ admits an extremum at $\phi = \phi_1$ with $V[\phi_1] \neq 0$, a de Sitter space solution trivially exists at $\xi = 0$.

$$\phi = \phi_1$$

$$R_{\mu\nu} = 3H^2 g_{\mu\nu} \,,\; H^2 = \frac{8\pi G}{3} V[\phi_1] \quad . \tag{23}$$

This type of (effective) potentials naturally arise in simple models through quantum corrections and finite temperature corrections. The solution is classicaly stable, if ϕ_1 is a minimum of $V[\phi]$.

3-B. Non-trivial solution

Non-vanishing ξ can lead to highly non-trivial solutions. Let us assume that 3-space is homogeneous and isotropic, having a line element (13). Accordingly the scalar field depends on only t ($dt = a\,d\eta$): $\phi = \phi(t)$. The scalar field equation is

$$\ddot{\phi} + 3\frac{\dot{a}}{a}\dot{\phi} + V'[\phi] + \frac{6}{a^2}(a\ddot{a} + \dot{a}^2 + 1)\phi = 0 \quad , \tag{24}$$

where the dot means d/dt. The t-t component of the Einstein equations is

$$(1-8\pi G\xi\phi^2)(\dot{a}^2 + 1) = \frac{8\pi G}{3} a^2 \{\frac{1}{2}\dot{\phi}^2 + V[\phi] + 6\xi\frac{\dot{a}}{a}\phi\dot{\phi}\} \quad . \tag{25}$$

Other components of the Einstein equations are either identities or trivial consequences of these equations unless $a(t)$ = const.

We seek for a self-consistent de Sitter space solution with $a(t) = H^{-1}\cosh Ht \equiv a_o(t)$, assuming that $\phi(t)$ = const. $\equiv \phi_o$.[4] Then Eqs. (24) and (25) yield

$$V'[\phi_o] + 12\xi H^2\phi_o = 0,$$
$$(1-8\pi G\xi\phi_o^2)H^2 = \frac{8\pi G}{3}V[\phi_o] \quad . \tag{26}$$

A solution may or may not exist, depending on $V[\phi]$.

To make it concrete, let us consider a potential

$$V[\phi] = \frac{1}{2}m^2\phi^2 + \frac{\lambda}{4!}\phi^4. \tag{27}$$

It immediately follows from Eq. (26) that

$$\phi_o^2 = -m^2/(8\pi G\xi m^2 + \frac{\lambda}{3!}) \quad ,$$
$$H^2 = \frac{2}{3}\pi G m^2\phi_o^2 \quad . \tag{28}$$

It is clear that a solution exists only if

$$m^2 > 0, \; 8\pi G\xi m^2 + \frac{\lambda}{3!} < 0 \quad . \tag{29}$$

At $\xi = 0$ and $\lambda < 0$, ϕ is on top of the potential so that the solution is classically unstable. We have to find under what conditions the solution (28) is classically stable. For small fluctuations $\phi_1 = \phi - \phi_o$ and $a_1 = a - a_o$. Eqs. (24) and (25) reduce to

$$\ddot{\phi}_1 + 3\frac{\dot{a}_o}{a_o}\dot{\phi}_1 + A\phi_1 = 0$$
$$A = \frac{(1-K)V''[\phi_o] + (1-3K)12\xi H^2}{1+(6\xi-1)K}$$
$$K = 8\pi G\xi\phi_o^2$$
$$\frac{1}{a_o}(\dot{a}_1 - \frac{a_o}{\dot{a}_o}a_1) = \frac{K}{1-K}\frac{1}{\phi_o}(\dot{\phi}_1 - \frac{a_o}{\dot{a}_o}\phi_1) \quad . \tag{30}$$

The solution is classically stable if, and only if,

$$A > 0. \tag{31}$$

The two conditions (29) and (31) are simultaneously satisfied only if

$$m^2 > 0 \ , \ 0 < \xi < \frac{1}{6}$$

$$24\xi(3\xi-1) < \frac{\lambda}{4\pi}\frac{1}{Gm^2} < -12\xi \ . \qquad (32)$$

The solution is classically stable for negative λ! In a flat space negative λ means that the theory is ill defined because of its unbounded potential. In a curved space a negative λ does not necessarily mean the instability. Indeed, due to the coupling ($\xi \neq 0$) to gravity theory can make sense if proper conditions are met. Anyhow, energy density is not a good criterion for the stability in curved space. It can be checked that $V[\phi_o] < 0$ but $R > 0$ in the stable solution.

4. Exact solutions in the early universe

One has to ask if various exact classical solutions presented in the previous section are important in the early universe. The answer is not quite clear at the moment. I would rather list questions arising naturally from consideration of those solutions.

First of all, is the picture reasonable that the very early universe is dominated by classical coherent fields? In the standard inflationary universe scenario finite temperature ($T \neq 0$) effects and quantum corrections play a crucial role in changing the form of the effective potential to admit a semistable de Sitter space solution. But what is necessary in the inflationary universe scenario is inflation. Many other ways of getting inflation have been suggested by several authors. Starobinski showed that Casimir effects in curved space lead to a consistent de Sitter space solution to the Einstein equations in massless theory.[5] Brout et al. have argued that the universe and matter are created as a cooperative phenomenon of quantum mechanics and general relativity, the initial stage being in an approximate de Sitter phase.[6] They analysed a model of a free (λ=0), but very massive ($m \sim M_p$), scalar field. Vilenkin showed that the universe can be created from nothing through semiclassical tunneling.[7] Inflation naturally takes place in this picture, since the universe starts with a scalar field on a local maximum of a potential. There may be other ways of getting inflation. From this point of view it is worth examining the picture that in the inflation or pre-inflation period the universe is dominated by classical coherent fields with large space curvature.

The solution in a conformally invariant scalar theory, (6) and (8), is of great interest, since this theory is adopted in the new inflationary universe scenario. As was pointed out before, the negative energy density (9) of the solution does not necessarily mean that this solution is more favored than the flat space solution. More importantly, quantum effects are expected to change the behavior of the solution. In a flat space quantum radiative corrections induce symmetry breaking by the Coleman-Weinberg mechanism[8] so that the behavior of the solution (6) must be changed at least when curvature $a(t)^{-1}$ gets smaller than a typical symmetry breaking scale in a flat space ($\sim M_{GUT}$). To see what happens for $a(t)^{-1} > M_{GUT}$ one has to evaluate an effective action for a time dependent configuration in curved space consistently. Unfortunately, it is a very hard task. If a solution of a similar kind to (6) is realized at the quantum level, the monopole problem[9] would be trivially solved, since the universe is always in a broken phase, say, in the SU(3) x SU(2) x U(1) phase at the GUT scale. The problem pointed out by Breit, Gupta and Zaks and by Moss[10] may be also solved for the same reason.

The exact solution in the pure Yang-Mills system, (18) and (20), would be irrelevant in the early universe in its original form. It predicts too short lifetime of the universe and is unstable. However, if other matter fields are coupled to the gauge fields and other effects (quantum effects?) are taken into account, the configuration (18) or its modified version may be realized in a certain epoch of the very early universe. The solution (18) is quite unique in a closed universe.

The de Sitter space solution (28) and (32) raises many interesting questions. First of all, the solution exists in a simple scalar field system (27). The condition (32) can be met either if $|\lambda|/4\pi$ ($\lambda<0$) is very small, or if m^2 is reasonably large. Negative λ is attractive, since the theory is asymptotically free. Is the flat space stable for negative λ? In the presence of the coupling to gravity ($\xi\neq0$) the answer is not clear. At the classical level it is stable. A criterion for the stability at the quantum (semi-classical) level is the existence or non-existence of a bounce (instanton) solution in the Euclidean signature. Coleman and De Luccia's analysis[11] suggests that a bounce solution might exist only if $\lambda < \lambda_c(m^2,G,\xi) < 0$. Assuming that the theory makes sense for λ satisfying (32), one has to ask which state is more stable, the flat Minkowski space or the de Sitter space (28)? In the de Sitter space the effective gravitational constant $G_{eff} = G/(1-8\pi G\,\xi\phi_o^{\,2})$ is negative at λ in the range (32). The anti-gravity nature of the de Sitter space suggests that the de Sitter space is only semistable, eventually decaying into the flat Minkowski space. It may also explain the origin of the big bang.

If the solution (28) is realized, inflation takes place in a broken phase for a charged ϕ. The picture is similar to the chaotic inflation scenario by Linde.[12] The difference lies in the fact that (28) solves the scalar field equations and the Einstein equation.

5. Dynamical $O(\hbar)$ mass generation by finite curvature

Quantum field theory in curved space (in background gravitational fields) is not well defined beyond one loop [$O(\hbar)$] order, at the moment. There is no proper regularization method which can be used to all orders in arbitrary background gravitational fields without spoiling either gauge invariance or general coordinate invariance. Nevertheless we explore genuine quantum phenomena in curved space to see their physical consequences, which may be consistently done to the one loop order by using, for instance, the zeta function regularization.[13] No doubt the Hawking radiation coming from a black hole is one of such phenomena.[14]

Another phenomenon is a phase transition caused by finite curvature. It is well known that symmetry can be restored at high temperature. Here finite curvature effects replace finite temperature effects.

5-A. Gauge theory in de Sitter space

Gauge theory in de Sitter space has been extensively investigated partly because of high symmetry residing in de Sitter space, and partly because the universe is supposed to be in a de Sitter phase in the inflation period. In particular, Shore and Allen have examined Coleman-Weinberg gauge theory in de Sitter space to evaluate an effective potential at the one loop level.[15] In a flat space gauge symmetry is spontaneously broken by radiative corrections. Shore and Allen found that the symmetry is restored if curvature is sufficiently large, i.e., if $H > H_c$ where H is defined in Eq. (23). Though there is a subtle problem in their treatment of renormalization as is discussed below (in 5-C), the qualitative feature of a phase transition caused by large curvature is expected to remain true.

5-B. ϕ^4-theory in the Einstein static universe

Let us consider a conformally invariant scalar field theory (1) in the Einstein static universe in which 3-space is given by a three-sphere S^3. In this spacetime the energy spectrum of a free scalar field is discrete:

$$\omega^2 = \frac{n^2}{r^2} \, , \; n = 1,2,3,\ldots \quad . \tag{33}$$

Here r is a radius of S^3. The energy spectrum is changed by quantum corrections. In particular, for small fluctuations in the S-wave mode (n=1 mode in (33)), namely for $\phi(t,x) = v \cos \omega t$ (v = const.), a dispersion relation is changed to

$$\omega^2 = \frac{1}{r^2} + m_{eff}^2 \quad . \tag{34}$$

m_{eff} is called a dynamical mass.

m_{eff}^2 was first evaluated by Ford and Yoshimura.[16] We give an alternative derivation, which turns out to be very useful in discussing gauge theory. The energy spectrum is determined by pole positions of two-point Green's functions. In other words one has to evaluate the effective action Γ_{eff} for a configuration $\phi = v \cos \omega t$ to $O(v^2)$, finding the spectrum from positions of zeros.

In the zeta function regularization method one loop corrections are

$$\frac{1}{\text{vol}} \Gamma^{(1)} [v \cos \omega t] = \frac{1}{4} v^2 \cdot \frac{\lambda}{96\pi^2} \frac{1}{r^2} + O(v^4) \quad . \tag{35}$$

The total effective action is

$$\frac{1}{\text{vol}}\,\Gamma_{\text{eff}}\,[v\cos\,\omega t] = \frac{1}{4}v^2\left\{\omega^2 - \frac{1}{r^2} + \frac{\lambda}{96\pi^2}\frac{1}{r^2}\right\} + O(v^4), \tag{36}$$

from which it follows that

$$m^2_{\text{eff}} = -\frac{\lambda}{96\pi^2}\frac{1}{r^2} \quad . \tag{37}$$

m^2_{eff} is negative! Of course, it does not imply the instability of the vacuum, since the classical term in Eq. (34) always dominates over the quantum correction at least in perturbation theory.

5-C. Gauge theory in the Einstein static universe

The dynamical mass generation problem in SU(N) Coleman-Weinberg gauge theory in the Einstein static universe has been investigated by the author.[17] The Lagrangian is

$$\mathcal{L} = -\frac{1}{2}\,\text{Tr}\,F_{\mu\nu}F^{\mu\nu} + \text{Tr}\,D_\mu\Phi\,D^\mu\Phi + \xi R\,\text{Tr}\Phi^2$$
$$- \frac{a}{4}\,\text{Tr}\,\Phi^4 - \frac{b}{4}\,(\text{Tr}\,\Phi^2)^2 \quad ,$$
$$D_\mu\Phi = \partial_\mu\Phi - ig\,[A_\mu,\Phi] \quad . \tag{38}$$

Here a scalar field Φ is in the adjoint representation of the group. Dominant contributions to m^2_{eff} are expected to come from gauge interactions for two reasons. First, gravity couples to spins and higher spin particles usually yield dominant contributions in curved space. Secondly ϕ^4-interactions are supposed to be much smaller than gauge interactions in the Coleman-Weinberg theory.[8]

Several comments must be made before going into detailed calculations. First of all, regularization and renormalization must be carefully handled. We use zeta function regularization. Dimensional regularization[18] is ambiguous in curved space, in which results generally depend on what kind of extra dimensions one adds to define regularized quantitie We will also see that renormalization is non-trivial in gauge theory even at the one loop level. The effective action must be evaluated for a time dependent configuration to know a wave function renormalization constant. Secondly, it must be kept in mind that the effective action itself is gauge dependent, though physical quantities like m^2_{eff} in (34) are gauge independent. Due caution is necessary in discussing global behavior of the effective potential.

We evaluate the effective action for a configuration

$$(\Phi_o)_{jk} = \delta_{jk}\,v_j\,\cos\,\omega t \;. \qquad \Sigma\,v_j = 0 \quad . \tag{39}$$

Calculations get simplified in the improved 't Hooft-Feynman gauge, in which

$$<A_o> = 0$$

$$\mathcal{L}_{\text{gauge fixing}} = - \text{Tr}\, \{F - <F>\}^2$$

$$F[A,\Phi] = A^{\mu}{}_{\mu} + ig\, [\Phi_o, \Phi] \ . \tag{40}$$

Here <f> denotes the S-wave component of f: $<f> = (2\pi^2 r^3)^{-1} \int d^3x \sqrt{-g}\, f$. One loop corrections in zeta function regularization are

$$\frac{1}{\text{vol}}\, \Gamma^{(1)}[\Phi_o] = \Sigma \frac{1}{2} v_j{}^2 \Big[\frac{1}{r^2} \frac{g^2 N}{4\pi^2} \{\ln \mu^2 r^2 - K(1) + \frac{3}{4} + \ln 2\pi\}$$

$$- \omega^2 \frac{g^2 N}{4\pi^2} \{\ln \mu^2 r^2 - K(\omega r) + \ln 2\pi\}\Big] + O(v^4) \ . \tag{41}$$

Here $K(\omega r)$ is a regular function of ωr with $K(1) = \ln 4 - 2\gamma - \frac{1}{2}$. μ is a dimensional parameter arising in the course of regularization.

The effective action at the tree level with counter terms is given by

$$\frac{1}{\text{vol}}\, \Gamma^{(\text{tree})}[\Phi_o] = \Sigma \frac{1}{2} v_j{}^2 \cdot Z\, (\omega^2 - Z_\xi \frac{1}{r^2}) + O(v^4) \ . \tag{42}$$

Z is a wave function renormalization constant, whereas Z_ξ is a renormalization constant associated with the coupling ξ in $\frac{1}{2}\xi R\phi^2$. [$\xi = \frac{1}{6}$ in the case at hand.] Comparing (41) with (42), one finds that

$$Z = 1 + \frac{g^2 N}{4\pi^2}\{\ln \mu^2 r^2 - K(\omega r = 1) + \ln \omega 2\}$$

$$Z_\xi = 1 \tag{43}$$

remove all μ-dependence in (41). Here we have normalized wave functions at $\omega^2 = 1/r^2$. [They cannot be normalized at $\omega^2 = 0$ in massless theory.] The total effective action is

$$\frac{1}{\text{vol}}\, \Gamma_{\text{eff}}[\phi_o] = \Sigma \frac{1}{2} v_j{}^2 \Big[\omega^2 (1 + \frac{g^2 N}{4\pi^2} \{K(\omega r) - K(1)\})$$

$$- \frac{1}{r^2}(1 - \frac{3g^2 N}{16\pi^2}) \Big] + O(v^4) \ . \tag{44}$$

The zero of (44) may be evaluated in a power series in g^2

$$\omega^2 = \frac{1}{r^2}\left\{1 - \frac{3g^2 N}{16\pi^2} + O(g^4)\right\} ,$$

$$m^2_{\text{eff}} = - \frac{3g^2 N}{16\pi^2} \frac{1}{r^2} \ . \tag{45}$$

Again m^2_{eff} is negative. The gauge field contribution (45) is quite large compared with contributions from ϕ^4-couplings, (37). Note that wave function renormalization is crucial to obtain (45). Computations of the effective potential for constant (time-independent) configurations alone cannot determine Z and accordingly m^2_{eff}.

The advantage of the improved 't Hooft-Feynman gauge (40) becomes clear in evaluating the effective potential. With the same normalization conditions as those used in (43) and (44) one finds for $(\Phi)_{ij} = \delta_{ij}\,\phi n_j$ $(\Sigma n_j = 0,\ \Sigma n_j^{\,2} = \tfrac{1}{2})$

$$V_{eff} = \frac{1}{2}\frac{1}{r^2}\phi^2 + A\phi^4(\ln\frac{\phi^2}{\sigma^2} - \frac{1}{2}) + \frac{1}{2\pi^2 r^4}\sum_{j<k} G(\kappa^2_{jk})$$

$$A = \frac{3g^4}{32\pi^2}\sum_{j<k}(n_j - n_k)^4 \quad ,$$

$$\kappa^2_{jk} = g^2 r^2 \phi^2 (n_j - n_k)^2 \quad . \tag{46}$$

In the flat space $(r\to\infty)$ V_{eff} has a minimum at $\phi = \sigma$. $G(\kappa^2)$ in the last term represents finite curvature effects of $O(\hbar)$ and is given by

$$G(\kappa^2) = 6\int_\kappa^\infty dx\,\frac{x^2(x^2-\kappa^2)^{\frac{1}{2}}}{e^{2\pi x}-1} + 4\int_\kappa^\infty dx\,\frac{(x^2-\kappa^2)^{\frac{1}{2}}}{e^{2\pi x}-1}$$

$$+ \kappa - (1+\kappa^2)^{\frac{1}{2}} + \kappa^2(\frac{1}{2}\ln\kappa^2 + \frac{3}{4} - \ln 2) \quad , \tag{47}$$

and has asymptotic behavior

$$\lim_{\kappa^2\to 0} G(\kappa^2) = \text{const.} - \frac{3}{8}\kappa^2$$

$$\lim_{\kappa^2\to\infty} G(\kappa^2) = \frac{1}{2}\kappa^2\ln\kappa^2 + (\frac{3}{4} - \ln 2)\,\kappa^2 + \text{const.} \tag{48}$$

If one calculates V_{eff} in the standard 't Hooft-Feynman gauge, in which no special case is taken for S-wave components, one would find extra contributions of the form $gr\phi|n_j - n_k|$. This singular behavior of V_{eff} for small ϕ has nothing to do with physics, being a gauge aftifact conspiring with infrared singularity.

Since V_{eff} (46) is free from artificial singularities, one may analyse the phase transition problem by investigating global behavior of V_{eff}. At $r^{-1} \ll \sigma$ $(r^{-1} \gg \sigma)$ the global minimum of V_{eff} is $\phi \sim \sigma$ $(\phi=0)$. As curvature of S^3 is increased, a phase transition takes place at $r^{-1}\sim\sigma$. The transition is of the first order at least for weak couplings.

Finally I would like to stress again the importance of wave function renormalization to obtain physical results. In the calculations of the effective potential in Coleman-Weinberg gauge theory in de Sitter space both Shore and Allen did not carry out wave function renormalization.[15] Consequently, their final results contain $\ln \mu r$ dependence. Note that once renormalization is properly carried out, $\ln \mu r$ dependence has completely disappeared from our final expressions (44) and (46). It seems that the de Sitter space calculations need to be re-examined.

6. Summary

I have discussed two aspects in gauge theory in curved space:

exact classical solutions and dynamical mass generation. I do not think that non-Abelian gauge theory in curved space has been fully understood. My analyses indicate that non-Abelian gauge theory in curved space not only involves many technical problems which need to be clarified further, but also leads to conceptually new phenomena which might have great relevance in the very early universe. The unique history of the universe might have originated from the intrinsic nonlinearity residing in non-Abelian gauge theory and Einstein's theory of gravity.

Acknowledgements

I am grateful to many participants in the workshop, particularly B. de Witt, J. Primack, and C. Wolf, for many comments and discussions. This work is supported in part by the U. S. Department of Energy under Contract No. EY-76-C-92-3071.

References

[1] A. H. Guth, Phys. Rev. D23, (1981) 347;
A. H. Guth and E. Weinberg, Phys. Rev. D23 (1981) 876;
A. D. Linde, Phys. Lett. 108B (1982) 389; ibid. 114B (1982) 431;
A. Albrecht and P. J. Steinhardt, Phys. Rev. Lett. 48 (1982) 1220.

[2] A. A. Grib, V. M. Mostepanenko, and V. M. Frolov, Theo. Mat. Fiz. 33 (1977) 42 [Theor. Math. Phys. 33 (1977) 869]; Phys. Lett. 65A (1978) 282;
A. A. Grib and V. M. Mostepanenko, Pis'ma Zh. Eksp. Teor. Fiz. 25 (1977) 302 [JETP Lett. 25 (1977) 277].

[3] Y. Hosotani, Univ. of Pennsylvania preprint, UPR-0253T (1984).

[4] Y. Hosotani, in preparation.

[5] A. A. Starobinsky, Phys. Lett. 91B (1980) 99.

[6] R. Brout, F. Englert and E. Gunzig, Ann. Phys. 115 (1978) 78;
R. Brout, F. Englert and P. Spindel, Phys. Rev. Lett. 43 (1979) 417;
R. Brout et al., Nucl. Phys. B170 FS1 (1980) 228.

[7] A. Vilenkin, Phys. Lett. 117B (1982) 25; Phys. Rev. D27 (1983) 2848.

[8] S. Coleman and E. Weinberg, Phys. Rev. D7 (1973) 1888.

[9] J. P. Preskill, Phys. Rev. Lett. 43 (1979) 1365.

[10] J. D. Breit, S. Gupta and A. Zaks, Phys. Rev. Lett. 51 (1983) 1007;
J. G. Moss, Phys. Lett. 128B (1983) 385.

[11] S. Coleman and F. De Luccia, Phys. Rev. D21 (1980) 3305.

[12] A. D. Linde, Phys. Lett. 129B (1983) 177.

[13] S. W. Hawking, Comm. Math. Phys. 55 (1977) 133.

[14] S. W. Hawking, Comm. Math. Phys. 43 (1975) 199.

[15] G. M. Shore, Ann. Phys. 128 (1979) 376;
B. Allen, Nucl. Phys. B226 (1983) 228; Univ. of California preprint, UCSB TH-1 (1984).

[16] L. H. Ford and T. Yoshimura, Phys. Lett. 70A (1979) 89;
D. J. Toms, Phys. Rev. D21 (1980) 2805;
L. H. Ford, Phys. Rev. D22 (1980) 3003;
G. Kennedy, Phys. Rev. D23 (1980) 2884;
G. Denardo and E. Spallucci, Nuovo Cimento 64A (1981) 27;
D. J. O'Connor, B. L. Hu, and T. C. Shen, Phys. Lett. 130B (1983) 31.

17] Y. Hosotani, Univ. of Pennsylvania preprint, UPR-0240T, to appear in Phys. Rev. D.
18] G. 't Hooft and M. Veltman, Nucl. Phys. B44 (1972) 189.

SPONTANEOUS QUANTUM COMPACTIFICATION IN SEVENTEEN DIMENSIONS†

Alan Chodos and Eric Myers

J.W. Gibbs Laboratory
Yale University
New Haven, Connecticut 06511

ABSTRACT

We consider general relativity with a cosmological term in D>4 space-time dimensions. We examine the possibility that the inclusion of one-loop quantum effects will allow compactification on a manifold of the form $M^4 \times S^N$ (N must be odd for technical reasons). The first candidate for such a phenomenon occurs for N=13. The associated grand unified gauge group is SO(14). The SO(14) gauge coupling constant is a predicted number in this theory.

The traditional appeal of Kaluza-Klein theories[1] has been more philosophical than phenomenological. The observation that the action for pure gravity in D>4 space-time dimensions:

$$S_D = \frac{-1}{16\pi G_D} \int d^D x \sqrt{|g|} \left(R - 2\Lambda \right) \tag{1}$$

becomes, after the process of "dimensional reduction"[2], the action for gravity coupled to a gauge theory in four space-time dimensions,

$$S_4 = \int d^4 x \sqrt{|\gamma|} \left\{ \frac{-1}{16\pi G_4} \left(R^{(4)}_{(\gamma)} - 2\Lambda \right) - \frac{1}{4} F^a_{\mu\nu} F^{\mu\nu a} + \text{scalar debris} \right\} \tag{2}$$

leads one to believe that the origin of the observed gauge theories is in higher dimensional gravity. As elegant and intriguing as this idea may be, there are formidable obstacles in really putting it to the test, stemming from the fact that the dimensional reduction takes place because the extra dimensions are assumed to form a compact manifold whose size is not much bigger than the Planck scale. It is hard to imagine how, with presently available energies, the extra dimension can be seen directly. Thus one is in danger of being able merely to reproduce, via dimensional reduction, a gauge theory that one could have written down in four dimensions to begin with.

One arena in which the extra dimensions might play a role is cosmology. In the very early universe, small length scales (i.e. high energies and temperatures) were important; furthermore, various cosmological solutions of higher-dimensional relativity[3] have the property that at sufficiently early times the extra dimensions were of comparable size to the dimensions that we now perceive. Studies have been done to relate higher dimensional theories to the problems of inflation and entropy production[4], and the question of possible time variation of fundamental "constants" in these models has been addressed[5].

A second possible way to explore the physical implications of Kaluza-Klein theories is to investigate their impact on grand unification. The gauge group that arises after dimensional reduction is the group of isometries of the internal manifold; furthermore, the value of the gauge coupling constant is determined by the ratio of the size of the internal dimensions to the Planck scale[6]. Thus if the dynamics is understood well enough to allow computation of both the symmetries and the size of the internal manifold, then both the gauge group and the magnitude of the gauge coupling constant will be predicted.

The purpose of this paper is to describe an example of such a process, where the dynamics is generated by one-loop quantum corrections to the equations of motion. The physics behind these corrections, and the method for computing them, has already been described at some length in the literature[7-10], so we shall omit most of the details here. Let us simply recall that the object of interest is the one-loop effective potential $V^{(1)}_{eff}$, which for a scalar field ϕ can be computed from

$$\exp\left[-V^{(1)}_{eff}(\phi_c)\Omega_M\right] = \int D\phi' \exp\left[-S_{cl}(\phi_c) - \tfrac{1}{2}(\phi', S''(\phi_c)\phi')\right]. \tag{3}$$

Here the ϕ' functional integration is to be performed in Euclidean space, and $S''(\phi_c)$ is the second variation of the classical action when expanded about ϕ_c. Ω_M is the (infinite) volume of four-dimensional Minkowski space. This formula is valid whether or not one is expanding about a classical solution, i.e. whether or not the first variation $S'(\phi_c)$ vanishes[11].

Performing the path integral, one obtains

$$V^{(1)}_{eff}(\phi_c) = \frac{1}{\Omega_M}\left[S_{cl}(\phi_c) + \tfrac{1}{2}\ln \mathrm{Det}\, S''(\phi_c)\right] . \tag{4}$$

The only significant modification of this formula in the case of a gauge theory (such as gravity) is that the DeWitt-Faddeev-Popov ghost matrix M_G must be taken into account:

$$V^{(1)}_{eff} = \frac{1}{\Omega_M}\left[S_{cl} + \tfrac{1}{2}\ln \mathrm{Det}\, S'' - \ln \mathrm{Det}\, M_G\right] . \tag{5}$$

The functional determinants in Eq. (5) are ill-defined; a convenient regularization is provided by the zeta-function technique[12]. For a given operator M one defines

$$\zeta_M(s) = \sum_i \lambda_i^{-s} \tag{6}$$

where one sums (or integrates) over all the eigenvalues λ_i of M. It then follows that

$$\frac{-d\zeta}{ds} = \sum_i \lambda_i^{-s} \ln \lambda_i \tag{7}$$

so that

$$\ln \operatorname{Det} M = \frac{-d\zeta}{ds}\bigg|_{s=0} \quad . \tag{8}$$

The strategy is then to compute $\zeta(s)$ via Eq. (6) for Re(s) sufficiently large, and to define the determinant by analytic continuation to s=0. We refer the reader to Reference (8) for further details in the case of higher-dimensional quantum gravity. Here we note only that to one-loop this procedure yields an unambiguous result if D is odd, but if D is even an extra (undetermined) renormalization is possible that destroys the predictive power of the calculation; we shall limit ourselves to D odd in the rest of this paper.

Of relevance to the present considerations is the work of Candelas and Weinberg[9], who considered a theory of gravity minimally coupled to f massless scalar fields in D dimensions:

$$S_{C-W} = \frac{1}{16\pi G_0} \int d^D x \sqrt{|g|} \left\{ R - 2\Lambda - \sum_{i=1}^{f} g^{\mu\nu} \nabla_\mu \phi_i \nabla_\nu \phi_i \right\} . \tag{9}$$

They assumed that the geometry is that of

$$M^4 \times S^N, \qquad D = N+4 \tag{10}$$

(here M^4 is Minkowski space) and that f is sufficiently large that the quantum fluctuations of the scalars dominates the contribution of the graviton. They showed that the content of the quantum corrections could be summarized by the requirement that for a solution of the form $M^4 \times S^N$ to exist at some radius $r=r_0$ for S^N, the equations

$$V_{eff}^{(1)}(r_0) = 0 \tag{11}$$

$$\left.\frac{dV_{eff}^{(1)}}{dr}\right|_{r=r_0} = 0 \tag{12}$$

should be obeyed.

The first equation says that there is no residual cosmological constant, i.e. that M^4 really is a solution, and the second demands that $r=r_0$ be an extremum of the potential. Here $V_{eff}^{(1)}$ is the total zero plus one-loop effective potential:

$$V_{eff}^{(1)} = V_{cl}(r) + V_Q(r) . \tag{13}$$

$V_{cl}(r)$ is computed simply by evaluating the classical action at the background geometry $M^4 \times S^N$:

$$V_{cl}(r) = \frac{-\Omega_N}{16\pi G_D}\left\{N(N-1)r^{N-2} - 2\Lambda r^N\right\} . \tag{14}$$

Here Ω_N is the volume of the unit N-sphere. The one-loop contribution $V_Q(r)$ is independent of the overall scale of the action, and therefore on dimensional grounds

$$V_Q(r) = \frac{f c_N}{r^4} \tag{15}$$

where C_N is a pure number. Since these are two equations to be satisfied, one hopes that a solution can be obtained provided one both adjusts r_0 and fine-tunes the bare cosmological constant Λ. Candelas and Weinberg showed that this was indeed the case, but that the C_N turned out to be unexpectedly small, of order 10^{-4} or 10^{-5}. Since the gauge coupling e is given by

$$e^2 = \left(\frac{N+1}{2}\right)\left[\frac{N(N-1) - 2\Lambda r_0^2}{f C_N}\right] \quad , \tag{16}$$

this means that a reasonable value of e^2 can be obtained only if f is chosen to be very large: $f=10^4$ or 10^5. They also computed the contribution of spinor fields, and again found that the C_N were very small.

With these results in mind, it becomes of some interest to know what the contribution of the gravitational field itself is to $V_Q(r)$ when the background is $M^4 \times S^N$. Unlike the 10^4 scalars of the Candelas-Weinberg computation, the graviton degrees of freedom are definitely present in any Kaluza-Klein theory (and their number is fixed by the theory). If scalars and spinors generate such a puny contribution to the dynamics, one may speculate that the graviton indeed may be the dominant effect.

The graviton is certainly the dominant effect in terms of computational complexity. Because of the product structure $M^4 \times S^N$, the degrees of freedom embodied in $g_{\mu\nu}$ have scalar, vector and tensor components when projected onto S^N. This results in a large number of different kinds of modes (about 10 different zeta-functions for S" and four for the ghost). These have all been described in the literature[8,10] and will not be recapitulated here.

In addition, there is a complication that arises in finding the possible values of r_0 and Λ. The formula for $V_{c\ell}(r)$, Eq. (14), remains the same, but because of the term $2\Lambda\sqrt{g}$, the quantum piece can now depend on Λ, and so will have the general form

$$V_Q(r) = \frac{F(y)}{(2\pi r)^4}$$

where y is the dimensionless variable $2\Lambda r^2$. Because F(y) can be (and in general is) a complicated function, the process of solving Eqs. (11) and (12) becomes more involved. It is convenient to recast Eqs. (11) and (12) in the form:

$$T(y_0) = \left[(N+2)N(N-1) - (N+4)y_0\right] F(y_0) - 2y_0\left[N(N-1) - y_0\right]\left.\frac{dF}{dy}\right|_{y=y_0} = 0 \tag{17}$$

and

$$\frac{r_0^{N+2}}{G_D} = \frac{16\pi}{(2\pi)^4\Omega_N}\left[\frac{F(y_0)}{N(N-1) - y_0}\right] \quad . \tag{18}$$

Here $y_0=2\Lambda r_0^2$.

The procedure to be followed then is:

(i) Determine F(y) via the zeta-function technique;

(ii) Use the equation $T(y_0)=0$ as an algebraic equation to fix the allowed values of y_0;

(iii) Determine the radius r_0 in terms of the D-dimensional Planck scale G_D by Eq. (18). The associated value of Λ is then trivially given by $\Lambda=(y_0/2r_0^2)$.

A simple consistency requirement is evident from Eq. (18). We assume that the D-dimensional theory describes gravity, not anti-gravity; hence $G_D>0$. The assumed metric makes sense only if $r_0>0$. Hence,

$$\frac{F(y_0)}{N(N-1)-y_0} > 0 \quad . \tag{19}$$

Put another way: if we find a root $T(y_0)=0$, then we require $F(y_0)>0$ if $y_0<N(N-1)$ and $F(y_0)<0$ if $y_0>N(N-1)$.

Let us list a few conditions on a solution that would seem to be necessary to make it physically acceptable:

(a) Ineq. (19) should be satisfied;

(b) r_0 should be a local minimum of V_{eff};

(c) r_0 should be the global minimum of V_{eff};

(d) r_0^{N+2}/G_D should be large enough that the loop expansion can be believed;

(e) V_{eff} should be real in a neighborhood of $r=r_0$.

Condition (a) has been discussed above. Condition (b) is intuitively obvious as a criterion for stability; however, it should be noted that deciding stability in general relativity can be difficult[21] and is certainly more involved than looking at V_{eff} as a function of a single variable r. Condition (c) is desirable to prevent tunnelling from $M^4\times S^N$ to another minimum. Condition (d) follows from the observation that the loop expansion is an expansion in inverse powers of the overall scale of the classical action, hence in powers of G_D, hence on dimensional grounds in powers of G_D/r_0^{N+2}. It must be admitted, however, that we have no way of knowing whether the true expansion parameter is G_D/r_0^{N+2} or $G_D/(2\pi r_0)^{N+2}$ or something else; hence the true meaning of "large enough" in condition (d) is not very precise.

Condition (e) will be dealt with further below, after we have discussed our results.

The case N=1 is actually much simpler than the others, in that many of the modes are not present. It was discussed fully in Ref. (8), where a solution was found that violated condition (b), i.e. $r=r_0$ was a maximum of V_{eff}.

For N=3,5,7,9,11 we find no solutions at all which satisfy condition (a). (Here we disregard any possible imaginary part, i.e. we look for solutions of Eqs. (17) and (18) with F replaced by Re(F).) Representative cases (N=5 and N=7) are shown in Figs. 1 and 2. The case N=5 is noteworthy because there is "almost" a root in the sense that T(y)=0 at a point y_0 between two nearby zeros y_1 and y_2 of F(y). The case N=7 is worth examining because of all the attention the seven sphere has received recently in the literature. Our result is that at least for pure gravity compactification does not take place at the one-loop level. There is a region, $37.2 < y < 42$, in which Re(F) has the correct sign, but T(y) does not have any zeros in this region.

The first N for which condition (a) is met is N=13, i.e. a total of 17 space-time dimensions. The relevant T(y) and F(y) are shown in Fig. 3. A zero of T(y) occurs in a region to the right of y=156 where F(y) is negative. The solution has $y_0=166.67$; $r_0/G_{17}^{1/15}=1.304$ and $\Lambda G_D^{2/15}=49$. The gauge coupling constant is

$$\alpha = \frac{e^2}{4\pi} = \frac{7}{2}(2\pi)^3\left[\frac{156-y_0}{F(y_0)}\right] = \frac{1}{16.07}\ . \tag{20}$$

The isometry group of S^{13} is SO(14). Thus on the basis of this solution we are led to predict SO(14) as the grand unification group. This group has already appeared in the literature[13] in the context of a viable candidate for unification rather than as the prediction of an underlying theory.

The significance of the numbers quoted above cannot really be assessed without renormalization. At the classical level, the D-dimensional and 4-dimensional gravitational constants are related by

$$G_4 = \frac{G_D}{r_0^N \Omega_N} \tag{21}$$

where Ω_N is the volume of the unit N-sphere. If we put in the numbers, we find that

$$\frac{r_0}{G_4^{1/2}} = 21.21 \tag{22}$$

Thus the compactification scale is 21 times the Planck scale. But the identification of $G_4^{1/2}$ with the Planck length $G^{1/2}$ is only valid classically; there will be one-loop corrections which will make G_4 different from G. It is even possible that the one-loop corrections will reverse the sign of G, thereby rendering the solution physically unacceptable[14].

In addition, the coupling constant α must be renormalized in two separate ways. Just as for G, the one-loop corrections must be evaluated. If this renormalization changes the sign of α, the solution will be unphysical. If not, the one-loop corrected α must still be renormalized from the Kaluza-Klein scale r_0 down to laboratory energies for comparison with experiment. If, in fact, the one-loop corrections do not drastically alter the properties of the solution, this will be evidence in favor of the hypothesis that condition (d) is met, i.e. that higher-order loop effects will be small.

We have provided illustrations of V_{eff} for the appropriate value of Λ in Figs. 4 and 5. As the reader can see, the solution represents a local but not a global minimum of the potential. Thus there is the possibility that tunnelling processes may render our solution unstable.

A more serious threat of instability comes from the fact that condition (e) is not met, i.e. there is an imaginary part to V_{eff}. This problem is generic in the sense that for any odd $N \gtrsim 3$ the imaginary part vanishes for one and only one value of y. Why this is so is easy to see. A typical zeta function has the form

$$\zeta_i(s) = \int \frac{d^4k}{(2\pi)^4} \sum_{l=l_0(i)}^{\infty} d_i(l,N) \left[k^2 + \frac{l^2 + l(N-1) + c_i(y)}{r^2} \right]^{-s}. \tag{23}$$

The integral over k arises because M^4 contributes a continuous part to the spectrum. The sum on ℓ comes from the modes on S^N; the sum starts at $\ell_0(i)=0$ for scalar modes, $\ell_0(i)=1$ for vectors and $\ell_0(i)=2$ for tensors. The factor $d_i(\ell)$ counts the degeneracy of each mode. The term $c_i(y)$ varies with each type of mode, being always of the form

$$c_i(y) = \alpha_i y + \beta_i(N) . \tag{24}$$

For the ghost modes, $\alpha_i=0$ (i.e. the ghost is independent of y) whereas for the modes coming from S" $\alpha_i=-1$.

The k integral can be explicitly performed:

$$\zeta_i(s) = \frac{\pi^2 r^{2s}}{(2\pi r)^4} \frac{\Gamma(s-2)}{\Gamma(s)} \sum_{\ell=\ell_0(i)}^{\infty} d_i(\ell,N) \left[K_i(N,\ell,y)\right]^{2-s} \tag{25}$$

where $K_i(N,\ell,y) \equiv \ell^2+\ell(N-1)+c_i(y)$.

Now

$$\ln \operatorname{Det} M_i \sim \left. \frac{-d\zeta_i}{ds} \right|_{s=0} . \tag{26}$$

The imaginary part comes from differentiating the factor K_i^{2-s} whenever $K_i(N,\ell,y)<0$ because one will then obtain $\operatorname{Im}\ell nK_i = \pi$. Setting $\ell n\operatorname{Det} M_i=(F_i(y))/(2\pi r)^4$, we find

$$\left|\operatorname{Im} F_i\right| = \frac{\pi^3}{2} \sum_{\ell=\ell_0(i)}^{\ell_1(i)} d_i(\ell,N)\, K_i^2(N,\ell,y) \tag{27}$$

where $\ell_1(i)$ is the largest value of ℓ such that $K_i(N,\ell,y)<0$. ($\ell_1(i)$ always exists because K_i is a monotomically increasing function of ℓ.) The total imaginary part can then be computed from

$$\operatorname{Im} F(y) = \frac{1}{2} \sum_{i\in S''} \operatorname{Im} \ln \operatorname{Det} M_i(y) - \operatorname{Im} \ln \operatorname{Det} M_G . \tag{28}$$

The ghost contributes a y-independent number to Im F(y). For y sufficiently small, $y<y_T$, none of the $K_i(N,\ell,y)$, $i \in S"$, will be negative for any i or ℓ. Hence Im F(y)=const. For $y>y_T$, increasingly more of the S" modes will be negative, and will contribute to Im F(y) with the opposite sign (in our convention, negative) to that of the ghost piece. Thus Im F(y) is a monotonically decreasing function of y; for $y<y_T$ it is a positive constant, and for some unique $y_1>y_T$ the contribution from the S" modes will precisely cancel this constant and Im $F(y_1)$ will vanish. This behavior is illustrated in Fig. 6. (The full Im V_{eff} is plotted, so the functional dependence has an extra $1/r^4$ in it.)

As of this writing, we do not understand the proper interpretation of Im V_{eff}. The fact that the ghost contributes to it suggests that it has something to do with gauge invariance. The fact that it changes sign (see Fig. 6) is disturbing, because if one sign of Im V_{eff} represents exponential decay, the other represents exponential growth[15]. Arguments have been given[16] that Im V_{eff} is spurious and should be set to zero; certainly if this prescription proves to be fully justified, it would make life a lot easier.

In this paper we have given an example of spontaneous quantum compactification induced by pure Einstein gravity on the manifold $M^4 \times S^{13}$. If the universe is indeed sitting in the little dimple pictured in Fig. 4, then SO(14) is the grand unified gauge group and the value of the gauge coupling constant is computable and will be known once the renormalizations mentioned earlier are carried out.

In addition to these renormalizations, it will also be of interest to compute V_{eff} for $M^4 \times S^N$ with N>13, to see whether S^{13} stands alone or whether there are other such manifolds that permit quantum compactification. From a phenomenological viewpoint, S^{17} and S^{21} would seem to be particularly interesting[17] whereas S^{15} has certain intriguing mathematical properties[18]. A search for solutions on these manifolds is presently underway.

Of course, a space-time of seventeen dimensions is not compatible with any known version of supergravity. Since there are instances where quantum compactification in pure gravity fails by only a little bit (see e.g. Fig. 1), one might hope[19] that the inclusion of the gravitino and other matter fields predicted by supergravity will allow for compactification on some eleven-dimensional manifold[20] (or perhaps ten-dimensional if the ambiguity associated with even dimensions can be resolved).

REFERENCES

†Presented by Alan Chodos.

1. Th. Kaluza, Sitzungsber. Preuss. Akad. Wiss. Phys. Math. Kl. 966 (1921); O. Klein, Z. Phys. 37, 895 (1926).
2. B. DeWitt in "Dynamical Theory of Groups and Fields" (Gordon and Breach, New York, 1965); Y.M. Cho and P.G.O. Freund, Phys. Rev. D12, 1711 (1975); L.N. Chang, K.I, Macrae and F. Mansouri, ibid. 13, 235 (1976).
3. A. Chodos and S. Detweiler, Phys. Rev. D21, 2167 (1980); P.G.O. Freund, Nucl. Phys. B209, 146 (1982).
4. E. Alvarez and M. Belen-Gavela, Phys. Rev. Lett. 51, 931 (1983); D. Sahdev, Phys. Lett. 137B, 155 (1984); Q. Shafi and C. Wetterich, Phys. Lett. 129B, 387 (1983).
5. W. Marciano, Phys. Rev. Lett. 52, 489 (1984).
6. S. Weinberg, Phys. Lett. 125B, 265 (1983).
7. T. Appelquist and A. Chodos, Phys. Rev. Lett. 50, 144 (1983); Phys. Rev. D28, 772 (1983).
8. A. Chodos and E. Myers, Annals of Physics 156 (in press).
9. P. Candelas and S. Weinberg, Nucl. Phys. B237, 397 (1984). See also Ya.I. Kogan and N.A. Voronov, JETP Letters 38, 311 (262) (1983).
10. M.A. Rubin and C.R. Ordonez, Texas preprints UTTG-10-83 and UTTG-5-84; M.H. Sarmadi, "Spontaneous Compactification in Quantum Kaluza-Klein Theories", ICTP preprint IC/84/3 (January 1984).

11. R. Jackiw, Phys. Rev. D9, 1686 (1974).

12. J.S. Dowker and R. Critchley, Phys. Rev. D13, 3224 (1976); S.W. Hawking, Commun. Math. Phys. 55, 133 (1977).

13. G. Feldman and R. Holman, J. Phys. 9, 7 (1983); R. Holman, ibid 9, 35 (1983); K.S. Soh and S.K. Kim, Phys. Rev. D26, 340 (1982); K. Yamamoto, Phys. Lett. 120B, 157 (1983); G.L. Shaw and F. Daghighian, Phys. Rev. D26, 1798 (1982).

14. An example of this has been noted on S^{17} by Sarmadi, Ref. 10.

15. We thank E. Witten for this remark (private communication at "Funland", Rehoboth Beach, Delaware 1984).

16. M. Rubin (private communication); P. Candelas, M. Rubin and C.R. Ordonez (in preparation).

17. M. Gell-Mann, P. Ramond and R. Slansky, "Complex Spinors and Unified Theories", in Supergravity, P. van Nieuwenhuizen and D.Z. Freedman, eds., North Holland (1979), p. 315.

18. We thank F. Gürsey for this observation.

19. We thank P. Ramond for expressing this hope to us.

20. P. Candelas and D.J. Raine, "From Anti-de-Sitter Space to Minkowski Space via Vacuum Fluctuations", preprint (1983).

21. J.L. Friedman, Commun. Math. Phys. 63, 243 (1978).
W.H. Press and S.A. Teukolsky, Ap. J. 185, 649 (1973).
T.J.M. Zouros and D.M. Eardley, Ann. of Phys., 118, 139 (1979).

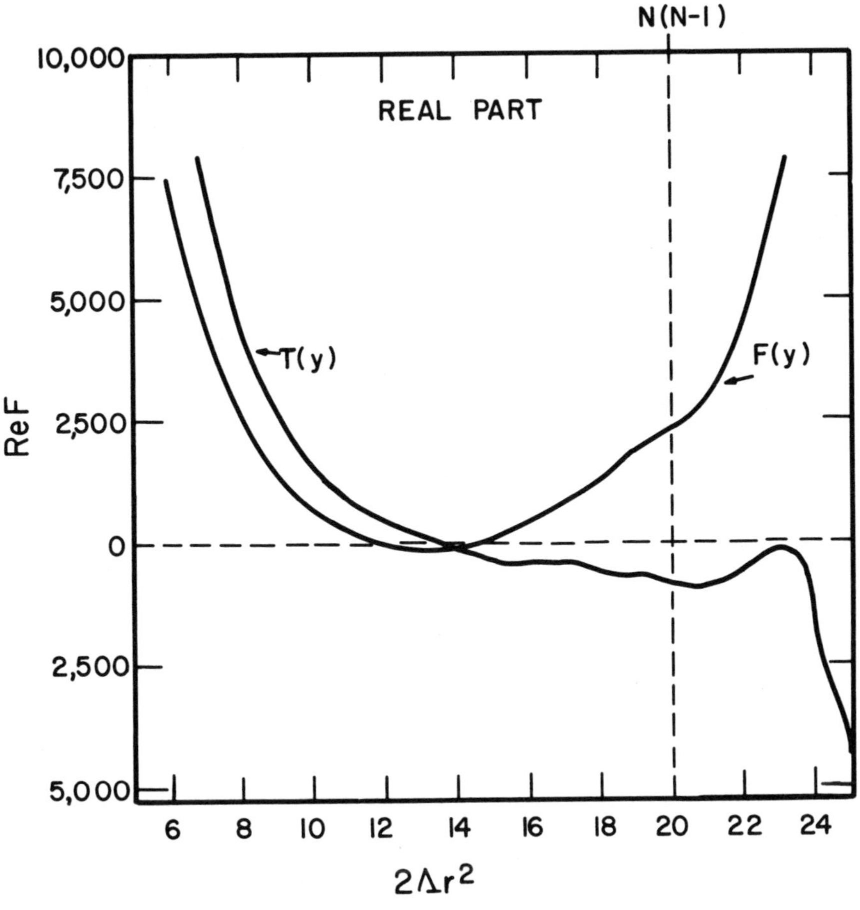

Figure 1. Real parts of T(y) and F(y) for N=5

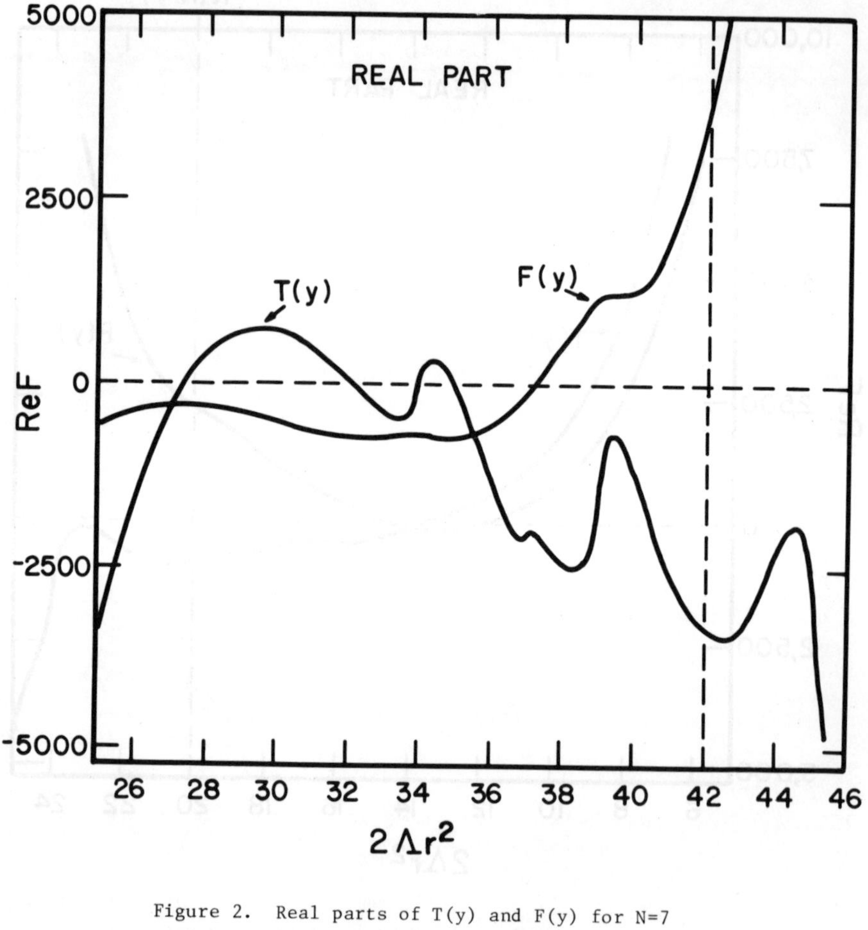

Figure 2. Real parts of T(y) and F(y) for N=7

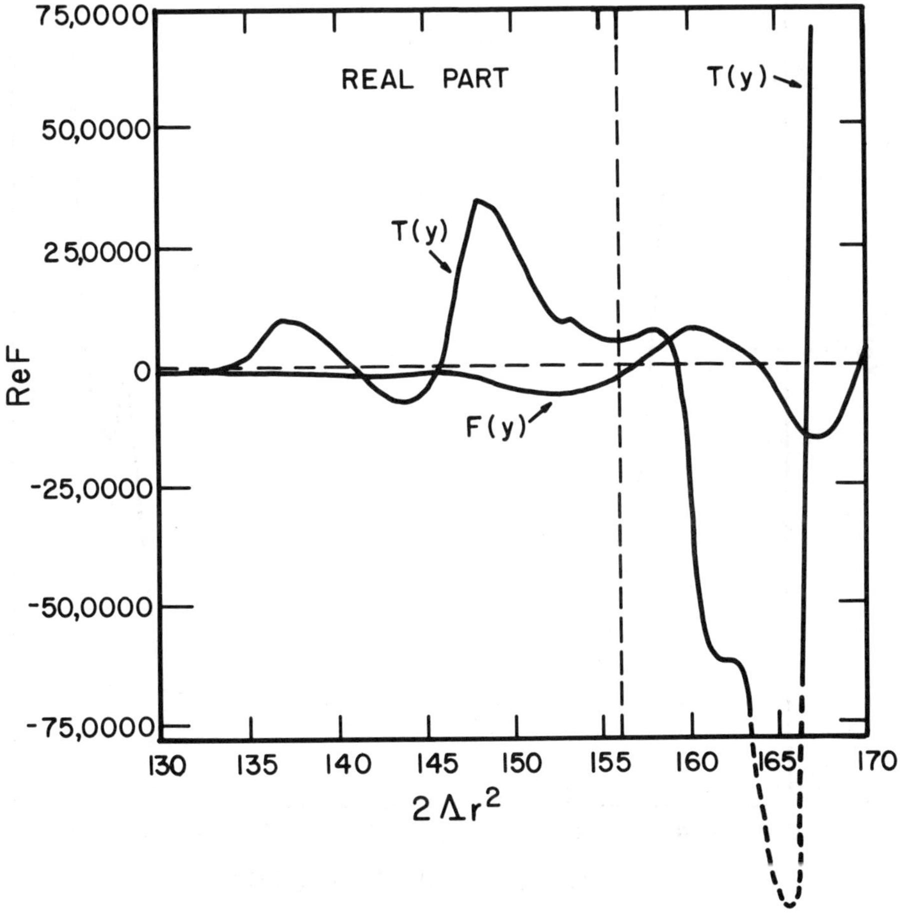

Figure 3. Real parts of T(y) and F(y) for N=13

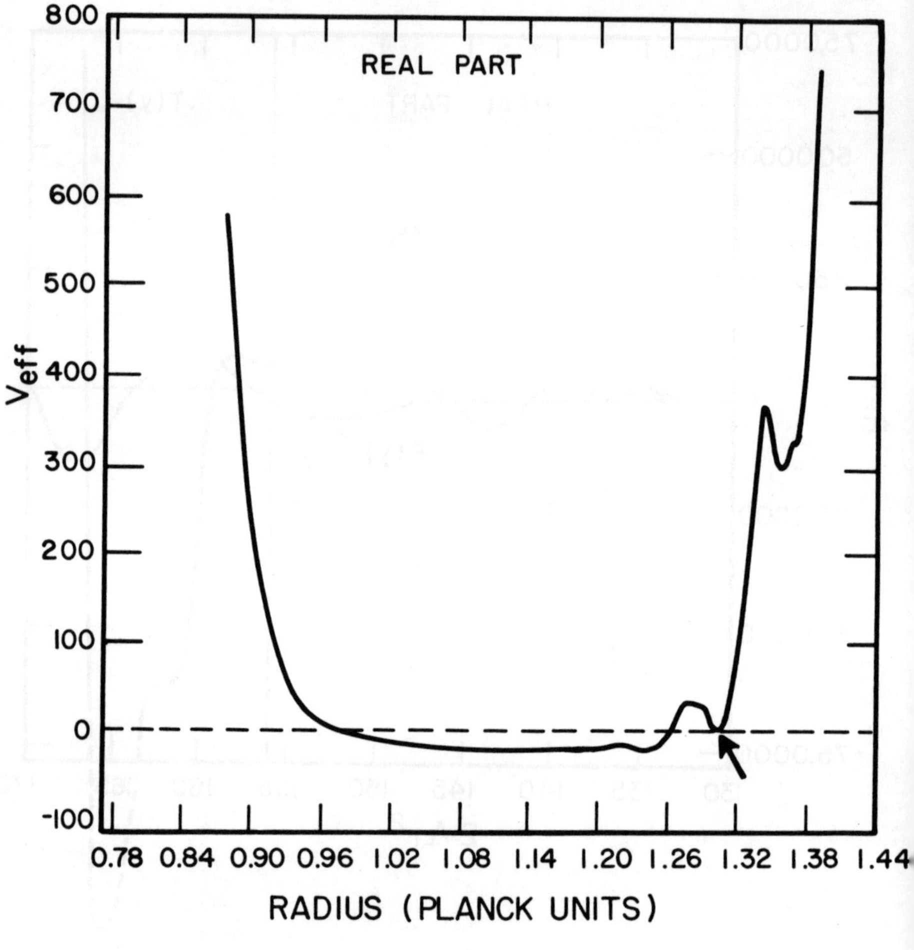

Figure 4. Effective potential -vs- radius for N=13

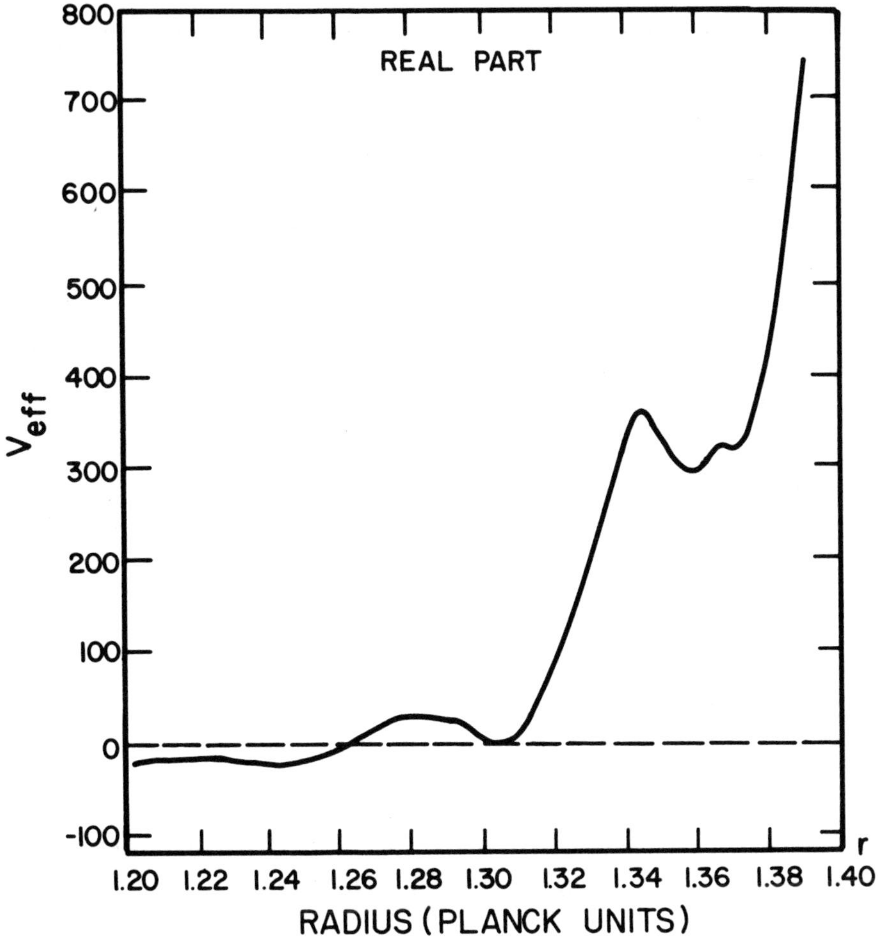

Figure 5. Effective potential -vs- radius for N=13

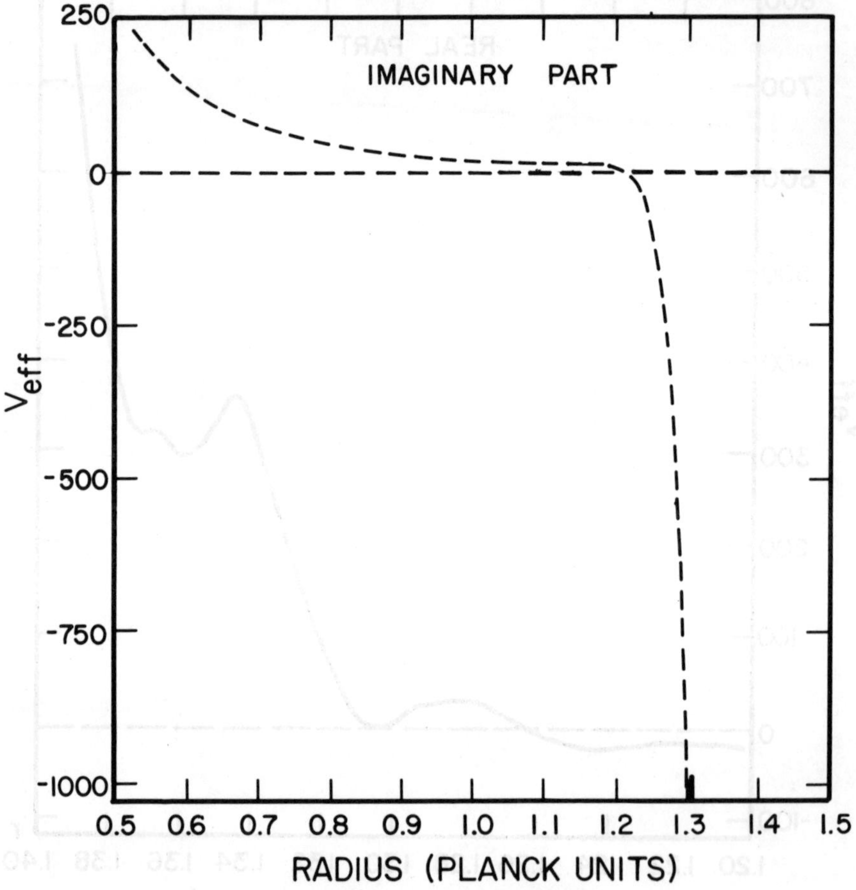

Figure 6. Imaginary part of the effective potential for N=13

GEOMETRY AS PARTICLES AND PARTICLES AS GEOMETRY

G.W. Gibbons

D.A.M.T.P
Silver Street
Cambridge CB3 9EW
UNITED KINGDOM

ABSTRACT

In this talk I describe some recent progress in implementing Einstein's Dream of constructing models for particles constructed solely from the geometry of spacetime, without singularities. I also mention the speculative possibility that there may be, at a fundamental level, no real distinction between "particles" and "geometry".

I wish to review in this talk some recent progress[1)] in the implementation of Einstein's Dream that one can build "particles" solely out of the geometry of spacetime. It turns out that progress in this direction seems to require another idea that, after some initial scepticism, he long advocated: the Kaluza-Klein theory. The other important ingredient - supersymmetry was not around in his day so one can only guess his reaction.

Einstein was unhappy because contemporary (classical) models of particles contained singularities - as for example in the Lorentz theory of the electron. He and others therefore looked in Pure Gravity for what we should nowadays call solitons, i.e. singularly free finite energy time independent solutions of the classical field equation $R_{\alpha\beta} = 0$. We now know due to a number of theorems[2)] that no such solutions exist. An essential ingredient in these proofs is the assumption that there are no horizons. Thus we may paraphrase these "no-go" theorems by the slogan, "No solitons without horizons". In fact, these results also hold in Einstein-Maxwell theory and probably Einstein-Yang-Mills theory.

If on the other hand, we allow horizons - and with them the inevitable singularities - then we have obvious contenders for solitons in the form of black holes. The singularities that these inevitably contain will be quite harmless, as far as the exterior of the horizon is concerned provided the Cosmic Censorship Hypothesis[3] holds. There is a unique two parameter family of black holes in pure Einstein theory parameterized by the mass M and angular momentum J[4]. In Einstein-Maxwell theory, one has a 4-parameter family parameterized by their electric charge Q and magnetic charge P[4]. Classically, none of these parameters is quantized.

The singularities will be hidden inside the event horizon if (setting J=0 for simplicity)

$$M \geqslant \left[(Q^2/\kappa^2) + (P^2/\kappa^2) \right]^{1/2} . \qquad (1)$$

This inequality is now known to be a consequence of supersymmetry[5]. It is also the case that equality (1) is, in the Einstein-Maxwell case, the condition that the horizon has a Hawking temperature of zero. This is important because a black hole with a non-zero Hawking temperature is unstable quantum mechanically against particle creation. Thus, since Q is a conserved central[6] charge, we discover that the only quantum mechanically stable black holes in Einstein-Maxwell theory are the supersymmetric extreme Reissner-Nordstrom holes[7].

These superholes cannot, however, really be thought of as "particles" in the usual sense of the word because the charge is not quantized and the mass essentially arbitrary. There appears, on simple dimensional grounds, to be no simple way of obtaining a quantization of electric of magnetic charge or mass unless one introduces a basic unit of charge e. This would seem to introduce an alien element into what is meant to be a purely geometric theory. Indeed to obtain stable black holes, we were forced to go to Einstein-Maxwell theory which can only (concievably) be thought of as geometrical if one thinks of it as N=2 ungauged supergravity. If one gauges the U(1), we are led to a cosmological constant $\Lambda = -3e^2/\kappa$ [8]. This means one does not have asymptotically flat spacetimes. One can discuss solitons in backgrounds which tend at large distances to anti-de Sitter spacetimes but as yet the physical significance is unclear. I shall merely note en passant that one finds Z_2 black hole monopoles which break supersymmetry.

There is, however, one very elegant way of obtaining a unit of charge in a theory whose basic Lagrangian does not contain one - that is to adopt the Kaluza-Klein programme[9] in which the radius of the 5th direction, R_k, and e are related by

$$e/\kappa = 4\pi/R_k . \tag{2}$$

Now quantum corrections to any gravitational soliton will, in general, diverge uncontrollably, since in general theories of gravity are non-renormalizable. Thus one is led to look at supergravity theories where these divergences are ameliorated and possibly eliminated because of cancellations between boson and fermion contributions. For our present purposes, the most attractive model to investigate is the N=8, d=5 supergravity model described by Cremmer[10]. This will give us a charge quantization and the matter fields will be unified with gravity via supersymmetry.

We can now anticipate that we will encounter magnetic monopoles whose magnetic charge P is quantized according to the Dirac relation

$$eP = 2\pi . \tag{3}$$

Indeed classical solutions corresponding to these monopoles have already been discussed by Cross, Perry and Sorkin[11]. The metric has the form

$$ds^2 = dt^2 - g_{\alpha\beta}dx^\alpha dx^\beta \tag{4}$$

where $g_{\alpha\beta}$ is the Riemannian (++++) metric of a multi Taub-NUT soliton of the Einstein equations $R_{\alpha\beta}=0$ [12]. The Riemann tensor $R_{\alpha\beta\gamma\delta}$ of $g_{\mu\nu}$ is self-dual

$$R_{\alpha\beta\gamma\delta} = 1/2[\varepsilon_{\alpha\beta\mu\nu}R^{\mu\nu}{}_{\gamma\sigma}] . \tag{5}$$

Cross and Perry show that the mass of these monopoles, M, is given by

$$M = P/2\kappa \quad (6)$$

The Cross-Perry-Sorkin monopoles are of course solutions of the N=8, d=5 supergravity model. One can therefore ask what supersymmetries do they posses. Now self-dual metrics admit 2 covariantly constant spinor fields which are chiral eigenstates. The gamma matrix corresponding to chirality is now γ^o (or γ "extra"). It is possible using these covariantly constant spinors to construct (for N=8) exactly half as many Killing spinors or supersymmetries as the vacuum state. This is in fact the maximum number of supersymmetries allowed for a system with central charges. One may further identify the magnetic charge $P/2\kappa$ with a central charge in 5 dimensions. Of course any electric charge Q would correspond to a component of momentum in the 5th direction equal to $Q/2\kappa$, by the usual Kaluza-Klein interpretation. It turns out that the Bogomolny inequality that follows from supersymmetry is now

$$M \geqslant \left[\frac{P^2}{2\kappa^2} + \frac{Q^2}{2\kappa^2} \right]^{1/2} . \quad (7)$$

At this point, it may prove useful to relate the Cross-Perry-Sorkin objects to the 4-dimensional black holes of the standard Kaluza-Klein theory[13]. These are similar to those of Einstein-Maxwell theory except that because of the scalar-vector couplings, the most, in general, possess a scalar charge Σ. This scalar charge is, however, not an independent parameter but should be thought of as a function of the 3 free parameters, the mass M, and electric charges P and Q. I am here considering only spherically symmetric holes. I believe that the introduction of angular momentum introduces no essentially new features but this has not been shown in detail. Cosmic censorship again imposes the inequalities which now take the form

$$M \geqslant \left[\frac{4\pi}{\kappa^2} \Sigma^2 + \frac{Q^2}{\kappa^2} + \frac{P^2}{\kappa^2} \right]^{1/2} , \quad (7)$$

$$QP \neq 0 \quad (8)$$

If equality holds in (7), we get zero temperature holes provided $QP \neq 0$. Conventional Kaluza-Klein theory does not possess a continuous duality invariance (unlike Einstein-Maxwell theory) however, the black hole solutions do have a discrete invariance under reversal of the scalar Σ and simultaneous interchange of electric and magnetic charges.

If one compares eqn. (7) with eqn. (8), one finds that being supersymmetric and having zero temperature are not the same thing in Kaluza-Klein theory (unlike Einstein-Maxwell theory). In fact, the only supersymmetric holes are the limiting cases

$$\text{A)}\quad M = P/2\kappa;\ Q = 0,\ \Sigma = \sqrt{3}\ KM/4\pi \tag{9}$$

$$\text{B)}\quad M = Q/2\kappa;\ P = 0,\ \Sigma = -\sqrt{3}\ KM/4\pi\ . \tag{10}$$

Case A corresponds to the Cross-Perry-Sorkin objects. These are singular considered as 4-dimensional objects but are regular in 5-dimensions. Case B is singular both in 4-dimensions and in 5-dimensions. However, this case is of considerable interest. All holes with P=0 may be obtained by boosting a 4-dimensional Schwarzschild hole into 5 dimensions. The limiting case is when the boost parameter is infinite and the objects move at the speed of light. Case B corresponds exactly to this case, provided a suitable rescaling is made.

In fact, Scherk showed some years ago[14)] that a massless scalar field in 5 dimensions should possess the quantum number given by (10), and hence should antigravitate. Later, I showed that singular black holes with these (9) properties also antigravitate[13)]. We see here the beginnings of an interesting parallelism between the fundamental fields of the theory and the particle-like solutions of the classical field equations. This brings me on to the question of supermultiplets. The Cross-Perry-Sorkin monopoles possess an interesting "zero mode" structure. The zero modes fall into supermultiplets or put more precisely using the zero modes one can see that the basic monopole would be, in the quantum theory, degenerate in mass with an entire supermultiplet of objects with spin up to spin 2, the spin arising from the zero energy fermion zero modes. This multiplet is a "short" multiplet because of the existence of central charges. In fact it is exactly the same multiplet that the massive states of the N=8, d=5 Kaluza-Klein theory fits into. When Scherk looked at antigravity, he realized that there exists an entire multiplet of antigravitating objects containing 256 states.

Is this a coincidence or is there something deeper here? One knows that in statistical mechanics and also in 2-dimensional field theories, there are surprising "dualities". The solitons of the Sine-Gordon model are the fermions of the Thirring model (10) for instance. Could something similar happen in Supergravity? Presumably at some level of approximation, there exists a field theory of the Cross-Perry-Sorkin monopoles. It must be very similar in structure to the theory of massive states of the Kaluza-Klein theory. Could it be, in some sense, identical? If this were true, then in some quantum mechanical sense, spacetime and the particles in it might be indistinguishable.

I realize that these are extremely speculative ideas. Much more needs to be done to see whether the speculation is even sensible let alone possibly true. Let me point out something in 3-dimensions which is curiously suggestive. Gravity in 3-dimensions is locally trivial, in that the metric is locally flat. However it does admit conical singularities supported on points in space or lines in spacetime which behave in many ways like particles[15)]. Now in Regge calculus[16)], spacetime is regarded as being made up (in 3-dimensions) of flat tetrahedral blocks. Curvature resides on lines. In fact these lines or bones are identical with the conical singularities I mentioned above. In the functional integral, one integrates over all possible Regge skeletons.

This is equivalent to integrating over all possible world lines and all possible masses for the "particles". Now the "particles" are <u>not</u> true solitons because the curvature blows up on the world line - equivalently one may say that there are sources on the world lines. Nevertheless, it is very suggestive that a fluctuating quantum spacetime may be considered as a gas of colliding particles in 3-dimensions - a theory in which the field equations lead to a trivial spacetime. Is it not possible that in 4 of 5 dimensions where the field equations are much less trivial, that something similar may happen?

How could one hope to pursue these ideas further? One obvious approach is to look at Regge Calculus in 4-dimensions or 5-dimensions. What do the monopoles look like in Regge calculus? Presumably they correspond to some sort of "defect". The study of topology in Regge calculus has only just begun. Nothing has been done so far on Supersymmetry in Regge calculus. Indeed no one has yet introduced fermions into Regge calculus. This will presumably be rather difficult. Another approach would be to look at gauged supergravity theories. As I mentioned above, one certainly has Z_2 monopoles then. However this will require a much deeper understanding of the cosmological constant problem in gauged supergravity. Finally, let me remark that important clues may come from the closely related question of the role of monopoles in Yang-Mills theory[18].

References

1) G.W. Gibbons and M.J. Perry, preprint.

2) R. Serini, Acad. Naz. Lince. Mem. Cl. Sci. Fiz. Nat. 27, 235 (1918); A. Einstein and W. Pauli, Ann. Math. 44, 131 (1943); A. Lichnerowicz, Théories Relativistes de la Gravitation a de l'Electromagnetisme, A. Mason et Cie, Paris 1955.

3) R. Penrose, Ann. N.Y. Acad. Sci. 224, 125 (1973).

4) D.C. Robinson, Phys. Rev. Lett., 34, 905 (1975). P.O. Mazur, J. Phys. A., 15, 3173 (1982).

5) G.W. Gibbons and C.M. Hall, Phys. Lett. 109B, 190 (1982); G.W. Gibbons, S.W. Hawking, G.T. Horowicz and M.J. Perry, C.M.P. 88, 295 (1983).

6) R. Haag, L. Lopusanski and M.F. Sohnius, Nucl. Phys. B88, 61 (1975).

7) G.W. Gibbons in Heisenberg Symposium, ed. P. Breitenlohner and H.P. Durr, Springer Lecture Notes in Physics, vol. 60 Springer Verlag 1982.

8) D.Z. Freedman and A. Das, Nucl. Phys. B120, 221 (1977).

9) T. Kaluza: Sitzungsbe. Preuss. Akad. Wiss. Math. Kl, 996 (1921) O. Klein: Z. Phys., 37, 895 (1926).

10) E. Cremmer in "Superspace and Supergravity", ed. S.W. Hawking, O.M. Rocek, Cambridge University Press, 1981.

11) D. Cross and M.J. Perry, Nucl. Phys. B226, 29 (1983); R. Sorkin, Phys. Rev. Lett. 51, 87 (1983).

12) S.W. Hawking, Phys. Lett. A, 60, 81 (1977).

13) P. Dobiasch and D. Maison, GRG 14, 231 (1982); A. Chodos and S. Detweiler, GRG 14, 879 (1982); G.W. Gibbons, Nucl. Phys. B207, 337 (1982).

14) J. Scherk in "Supergravity", eds. P. von Niewenhuizen and D.Z. Freedman, North Holland, 1979.

15) S. Deser, R. Jackiw and S. Templeton, Phys. Rev. Lett., 48, 975 (1982).

16) T. Regge, Nuovo Cimento 19 558 (1961)

17) J. Cheeger W. Muller and R. Schrader, C.M.P. 92 408 (1984); Humber and R.M. Williams - preprint.

18) D. Olive in N. Craigie, P. Goddard, W. Nahm eds. "Monopoles in Quantum Field Theory", World Scientific Publishing Co., 1982.

GRAVITATIONAL AND LORENTZ ANOMALIES

F. Langouche*), T. Schücker and R. Stora+)

CH-1211 Geneva 23
Switzerland

ABSTRACT

We define the anomalies for the symmetries of general relativity and describe the Adler-Bardeen class of solutions of the corresponding consistency equations in even dimensions by a purely algebraic algorithm.

Anomalies are said to occur when symmetries of a classical theory are broken by quantum corrections. A convenient way to describe a quantum theory is by its effective action Γ[1)]

$$\Gamma := S + \mathit{loops}$$

where S is the classical action. Since in general loop graphs have to be renormalized, Γ is defined only up to counter terms, local functionals in the fields Γ_{loc}. Suppose that our classical action is invariant under a continuous group with the Lie algebra $\mathfrak{g}$

$$W(g)\, S[\varphi_i] = 0 \qquad g \in \mathfrak{g} \tag{1}$$

$$W(g) = \sum_i \int dy\, W(g)\, \varphi_i(y) \frac{\delta}{\delta \varphi_i(y)} \tag{2}$$

with W(g) representing the Lie algebra on the fields ϕ_i. The anomaly $\mathfrak{a}(g)$ is a differential form of maximal degree on space-time defined by:

$$\int \mathfrak{a}(g) := W(g)\, \Gamma\,. \tag{3}$$

---

*) On leave of absence from K.U. Leuven, Onderzoeker IIKW, Belgium.

+) LAPP, Annecy, France.

It is naturally defined up to variations of local functionals and up to exact forms:

$$\mathcal{A}(g) \sim \mathcal{A}(g) + W(g)\Gamma_{loc} + d\chi \tag{4}$$

W(g) being a representation of $\mathfrak{g}$:

$$W(g)\,W(g') - W(g')\,W(g) = W([g,g']) \tag{5}$$

the anomaly satisfies the following consistency equations due to Wess and Zumino[2]:

$$W(g)\,\mathcal{A}(g') - W(g')\,\mathcal{A}(g) \sim \mathcal{A}([g,g']) \; . \tag{6}$$

Explicit calculations of the anomaly by Feynman graphs are generally very cumbersome. For gauge theories, however, there is a purely algebraic algorithm to find non-trival solutions of the consistency equations[3]. Furthermore, for renormalizable theories in four dimensions, all solutions are obtained in this way[4].

In the following we generalize this algorithm to the symmetries of general relativity[5].

The symmetries of general relativity

Consider the classical action of, say, a left-handed spin-½ field ψ in a gravitational field described by a co-moving frame e and a spin connection ω, both one-forms on a space-time manifold M of even dimensions 2n-2

$$S = \int_M d^{2n-2}x \; \det e \; \bar{\psi}\, e^{\mu}{}_{a}\, \gamma^{a}\left(\partial_\mu + \frac{1}{8}\omega_{\mu cd}[\gamma^c,\gamma^d] + \right.$$

$$\left. + \frac{1}{2} T^{\lambda}{}_{\mu\lambda}\right) \frac{1-\gamma_5}{2}\, \psi \; . \tag{7}$$

Note the torsion term needed to render the action hermitian. S is invariant under local Lorentz transformations and under diffeomorphisms. Let $\mathfrak{g}$ be their Lie algebra, sum of the Lorentz gauge algebra and the Lie algebra of vector fields on M. We denote their elements by Ω and ξ respectively:

$$W(\Omega)\psi = -\frac{1}{8}\Omega_{ab}[\gamma^a,\gamma^b]\,\psi \tag{8a}$$

$$W(\Omega)e = -\Omega e \tag{8b}$$

$$W(\Omega)\omega = d\Omega + [\omega, \Omega] \tag{8c}$$

$$W(\xi)\psi = -L_\xi \psi - \frac{1}{8} i_\xi \overset{\circ}{\omega}_{ab} [\gamma^a, \gamma^b] \psi \tag{9a}$$

$$W(\xi)e = -L_\xi e - (i_\xi \overset{\circ}{\omega}) e \tag{9b}$$

$$W(\xi)\omega = -L_\xi \omega + d i_\xi \overset{\circ}{\omega} + [\omega, i_\xi \overset{\circ}{\omega}] \tag{9c}$$

$\overset{\circ}{\omega}$ is a fixed (i.e., $W(\Omega)\overset{\circ}{\omega} = W(\xi)\overset{\circ}{\omega} = 0$) spin connection needed to patch these transformation laws together. The commutators of $\mathcal{G}$ follow:

$$[W(\Omega), W(\Omega')] = W([\Omega, \Omega']) \tag{10a}$$

$$[W(\xi), W(\xi')] = W([\xi, \xi']) - W(i_\xi i_{\xi'} \overset{\circ}{R}) \tag{10b}$$

$$[W(\xi), W(\Omega)] = W(L_\xi \Omega) + W([i_\xi \overset{\circ}{\omega}, \Omega]) \tag{10c}$$

with curvature defined as usual

$$\overset{\circ}{R} := d\overset{\circ}{\omega} + \frac{1}{2}[\overset{\circ}{\omega}, \overset{\circ}{\omega}] . \tag{11}$$

If M is parallelizable, e and ω are globally defined and we can put $\overset{\circ}{\omega} = 0$. Then $\mathcal{G}$ is just a semi-direct product. Changing $\overset{\circ}{\omega}$ does not change $\mathcal{G}$ but only amounts to a change of basis.

We now quantize ψ, keeping e and ω as classical (external) fields. Then the symmetry $\mathcal{G}$ will be broken leading to Lorentz and gravitational anomalies[6]:

$$\int \alpha(\Lambda, e, \omega) := W(\Lambda)\ \Gamma(e, \omega, \psi) \tag{12a}$$

$$\int \alpha(\xi, e, \omega) := W(\xi)\ \Gamma(e, \omega, \psi)\,. \tag{12b}$$

Note that ψ drops out, because it appears only in the classical action, which is invariant. The anomalies satisfy the consistency equations derived from (10):

$$W(\Lambda)\,\alpha(\Lambda') - W(\Lambda')\,\alpha(\Lambda) \sim \alpha([\Lambda, \Lambda']) \tag{13a}$$

$$W(\xi)\,\alpha(\xi') - W(\xi')\,\alpha(\xi) \sim \alpha([\xi, \xi']) - \alpha(-i_\xi i_{\xi'} \mathring{R}) \tag{13b}$$

$$W(\xi)\,\alpha(\Lambda) - W(\Lambda)\,\alpha(\xi) \sim \alpha(L_\xi \Lambda) + \alpha([-i_\xi \mathring{\omega}, \Lambda])\,. \tag{13c}$$

For algebraic convenience (and not for quantization purposes) we now go to the dual space of $\mathfrak{g}$.

Faddeev-Popov ghosts and Slavnov operator

For illustration, let us consider first the case of a finite dimensional Lie algebra $\mathfrak{g}$ with basis g_i and structure constants $f_{ij}{}^k$:

$$[g_i, g_j] =: f_{ij}{}^k g_k\,. \tag{14}$$

The dual basis consists of one-forms $\overset{*}{g}{}^i$ on $\mathfrak{g}$ defined by

$$\overset{*}{g}{}^i(g_j) = \delta^i{}_j\,. \tag{15}$$

We define the co-boundary (or Slavnov) operator s

$$s\overset{*}{g}^{k} := -\frac{1}{2} f_{ij}{}^{k}\, \overset{*}{g}^{i} \wedge \overset{*}{g}^{j} . \tag{16}$$

By transposing the Jacobi identity, one gets

$$s^{2} \overset{*}{g}^{k} = 0 . \tag{17}$$

The Maurer-Cartan form (or $\phi\pi$ ghost) is by definition:

$$\overset{*}{g} := \sum_{i} \mathfrak{g}_{i} \otimes g^{*i} \tag{18}$$

(fancy names for the identity mapping of $\mathfrak{g}$). We consider $\overset{*}{g}$ as one-form on $\mathfrak{g}$ with (non-transforming) values in $\mathfrak{g}$. As such, it is anticommuting and Eq. (16) can be written without reference to a particular basis:

$$s\overset{*}{g} = -\frac{1}{2} [\overset{*}{g}, \overset{*}{g}] . \tag{19}$$

Here and from now on the wedge symbol $\wedge$ is omitted. In the same spirit, a representation R of $\mathfrak{g}$ on V is considered as zero-form on $\mathfrak{g}$ with values in V and

$$sV := R(\overset{*}{g})\, V . \tag{20}$$

Transposing $[R(g), R(g')] = R([g,g'])$, one gets $s^2 V = 0$.

Now we return to our particular infinite dimensional Lie algebra (10). We define ghosts $\overset{*}{\Omega}$ and $\overset{*}{\xi}$ as projections on the Lorentz, respectively vector field parts of $\mathfrak{g}$. The additional operations d and i_ξ defined on $\mathfrak{g}$ can be transposed to its dual $\mathfrak{g}^*$ in such a way that $\overset{*}{d}$ anticommutes with s and acts as anti-derivation while $\overset{*}{i}_\xi$ now becomes a derivation. Then Eqs. (8b,c), (9b,c) and (10) are, with all the stars * dropped from now on, reduced to:

$$se = -\Omega e - L_{\xi} e - (i_{\xi} \mathring{\omega})\, e \tag{21a}$$

$$s\omega = -d\Omega - [\omega, \Omega] - L_{\xi}\omega - d i_{\xi}\mathring{\omega} - [\omega, i_{\xi}\mathring{\omega}] \tag{21b}$$

$$s\Omega = -\tfrac{1}{2}[\Omega,\Omega] - L_\xi \Omega - \tfrac{1}{2} i_\xi i_\xi \mathring{R} - [i_\xi \mathring{\omega}, \Omega] \qquad (21c)$$

$$s\xi = -\tfrac{1}{2}[\xi,\xi] \, . \qquad (21d)$$

Replacing Ω and ξ in (12) by their ghosts, we define

$$\alpha := \alpha(\Omega) + \alpha(\xi) \qquad (22)$$

a differential (2n-2)-form on M, which is at the same time a one-form on $\mathcal{G}$. It satisfies

$$\int_M \alpha = s\Gamma \, . \qquad (23)$$

The consistency equations (13) are now simply:

$$s\alpha \sim 0 \, . \qquad (24)$$

Solutions of the consistency equation

Let J be an n-linear symmetric, Lorentz invariant form, on the Lorentz algebra so(2n-3,1), for example in two dimensions (n = 2):

$$J(A,B) = \mathrm{tr}\, AB \qquad A,B \in so(1,1) \, . \qquad (25)$$

Such invariants only exist for even n. The following lemma is known as homotopy formula[7].

Let $\mathbb{W}_0$ and $\mathbb{W}_1$ be two connections and $\mathbb{d}$ a co-boundary operator, $\mathbb{d}^2 = 0$. Define an interpolating connection

$$\mathbb{W}_t := \mathbb{W}_0 + t(\mathbb{W}_1 - \mathbb{W}_0) \qquad t \in [0,1] \, . \qquad (26)$$

Let $\mathbb{R}_0$, $\mathbb{R}_1$ and $\mathbb{R}_t$ be the corresponding curvatures with respect to $\mathbb{d}$, e.g.

$$\mathbb{R}_0 := \mathbb{d}\, \mathbb{W}_0 + \tfrac{1}{2}[\mathbb{W}_0, \mathbb{W}_0] \, . \qquad (27a)$$

Then

$$\mathbb{J}(\mathbb{R}_1{}^n) - \mathbb{J}(\mathbb{R}_0{}^n) = \mathbb{d}\, Q \tag{28}$$

where

$$Q := n \int_0^1 dt \;\; \mathbb{J}(\mathbb{w}_1 - \mathbb{w}_0 \,,\, \mathbb{R}_t{}^{n-1}) \,. \tag{29}$$

We use this lemma by putting

$$\mathbb{w}_0 := \mathring{\omega} \tag{30a}$$

$$\mathbb{w}_1 := \omega + \Omega - i_\xi(\omega - \mathring{\omega}) \tag{30b}$$

$$\mathbb{d} := d + s \,. \tag{30c}$$

A straightforward calculation gives:

$$\mathbb{R}_1 = R - i_\xi R + \tfrac{1}{2}\, i_\xi\, i_\xi\, R \tag{31}$$

Now we decompose Q

$$Q = Q^0_{2n-1} + Q^1_{2n-2} + Q^2_{2n-3} + \ldots \tag{32}$$

where Q^j_i is an i-form on M and a j-form on $\mathfrak{g}$. We take the component of Eq. (28) which is a (2n-3)-form on M, 2-form on $\mathfrak{g}$ and get

$$\tfrac{1}{2}\, i_\xi\, i_\xi\, \mathbb{J}(R^n) = d\, Q^2_{2n-3} + s\, Q^1_{2n-2} \,. \tag{33}$$

But $\mathbb{J}(R^n)$ is a 2n-form over a (2n-2) dimensional manifold, hence zero and dQ^2_{2n-3} is exact. Therefore

$$\alpha = Q^1_{2n-2} \tag{34}$$

solves the consistency equation (24). We can calculate α and get:

$$\mathfrak{a}(\Omega) = n\int_0^1 dt\, J(\Omega, R_t^{\,n-1}) + $$
$$+\, n(n-1)\int_0^1 dt\, J(\omega - \overset{\circ}{\omega}, (t^2 - t)[\Omega, \omega - \overset{\circ}{\omega}], R_t^{\,n-2}) \qquad (35a)$$

$$\mathfrak{a}(\xi) = n(n-1)\int_0^1 dt\, J(\omega - \overset{\circ}{\omega}, (1-t)\, i_\xi \overset{\circ}{R}, R_t^{\,n-2}) \qquad (35b)$$

with

$$R_t := d\{\overset{\circ}{\omega} + t(\omega - \overset{\circ}{\omega})\} + \tfrac{1}{2}[\quad,\quad] . \qquad (36)$$

For example, in two dimensions where the Lorentz algebra is Abelian:

$$\mathfrak{a}(\Omega) = \mathrm{tr}\, \Omega\, d(\omega + \overset{\circ}{\omega}) \qquad (37a)$$

$$\mathfrak{a}(\xi) = \mathrm{tr}\,(\omega - \overset{\circ}{\omega})\, i_\xi\, d\overset{\circ}{\omega} \qquad (37b)$$

coinciding with explicit calculations of the spin-½ anomaly for $\overset{\circ}{\mathfrak{B}} = 0$[8]. On parallelizable space-times, one can choose $\overset{\circ}{\mathfrak{B}}$ flat

$$\overset{\circ}{R} = 0 \qquad (38)$$

in which case the gravitational anomalies vanish

$$\mathfrak{a}(\xi) = 0 . \qquad (39)$$

On the other hand, Bardeen and Zumino[9] have remarked that there is a local counter term Γ_{loc} such that $\mathfrak{a} + s\Gamma_{loc}$ is independent of Ω, which means vanishing Lorentz anomaly for this representative. The counter term Γ_{loc} is just the Wess-Zumino-Witten Lagrangian[2),10)] where the co-moving frame e considered as zero-form with values in the general linear group $G\ell(2n-2)$ plays the role of the Goldstone field[5)].

Other solutions?

It is not known whether all Lorentz and gravitational anomalies are of the above type. Alvarez-Gaumé and Witten have calculated explicitly the gravitational anomalies for spin-½, spin-3/2, and for the self-dual antisymmetric tensor[6]. Under some disguised form[5),9)], their results are all of the above type, with J depending on the propagating chiral field in a way which is now known to be related to the index of an elliptic operator in two more dimensions[11]. In particular, there are no anomalies in four dimensions.

References

1) See, for example: C. Nash - "Relativistic Quantum Fields", Academic Press (1978).

2) J. Wess and B. Zumino - Phys.Lett. 37B (1971) 95.

3) R. Stora - Lecture Notes, Cargèse 1976 in "New Developments in Quantum Field Theory and Statistical Mechanics", Eds. M. Lévy and P. Mitter, NATO ASI ser. B, Vol. 26, Plenum Press (1977);
B. Zumino - Lectures, Les Houches 1983 in Relativity Groups Topology II, Eds. B.S. De Witt and R. Stora, North Holland (1984), to appear;
R. Stora - Lectures, Cargèse 1983, Progress in Gauge Field Theory, Ed. H. Lehmann, to appear;
J. Mañez, R. Stora and B. Zumino - to be published.

4) C. Becchi, A. Rouet and R. Stora - in Field Theory, Quantization and Statistical Physics, Ed. E. Tirapegui, Reidel (1981), pp. 3-32.

5) F. Langouche, T. Schücker and R. Stora - CERN Preprint TH. 3898, LAPP-TH 10 (1984).

6) L. Alvarez-Gaumé and E. Witten - Nucl.Phys. B234 (1984) 269.

7) S. Kobayashi and K. Nomizu - "Foundations of Differential Geometry", Intersciences (1963);
S. Chern - "Complex Manifolds without Potential Theory", Springer (1979).

8) F. Langouche - CERN Preprint (1984).

9) W.A. Bardeen and B. Zumino - Preprint LBL17639, UCB-PTH 84-12, Fermilab PUB 84/38-T (1984).

10) E. Witten - Nucl.Phys. B223 (1983) 422.

11) O. Alvarez, I.M. Singer and B. Zumino - Comm.Math.Phys., to appear.

QUANTUM ORIGIN OF THE UNIVERSE[†]

A. Vilenkin

Department of Physics
Tufts University
Medford, MA 02155

[†]Following the speaker's instructions, no written version of this talk is included in these Proceedings.

The Editors

TOWARDS UNIFICATION OF ELEMENTARY PARTICLE PHYSICS AND COSMOLOGY IN 10-DIMENSIONS

George Chapline**
Lawrence Livermore National Laboratory
Livermore, CA 94550

Gary Gibbons
Department of Applied Mathematics and Theoretical Physics
Cambridge University
Cambridge, England

ABSTRACT

Ten-dimensions seems to be a unique setting for unifying at the classical level cosmology and elementary particle physics. Some interesting results along these lines are obtained starting with a Yang-Mills coupled to supergravity theory in 10-dimensions. However, further progress will require finding an underlying quantum theory.

One of the long held hopes of science is that some day a unified theory of the large scale structure of the universe and elementary particle physics would emerge. In a provocative paper Freund[1] suggested that such unification might be acheived by using the classical equations for 10- or 11-dimensional supergravity to describe cosmology. One hopes that elementary particle physics will emerge in this approach if the extra 6- or 7-dimensions are "spontaneously compactified" into a space with sufficiently small scale size. In particular, it has been eloquently argued[2] that 11-dimensional supergravity naturally leads to the compactification of 7-dimensions. Whether an acceptable phenomenology for elementary particle physics could emerge from compactification of 7-dimensions is unclear, however. For example, 11-dimensional supergravity is not chiral in <u>any</u> dimension; and therefore it is not clear how one would explain the parity violation observed in β-decay interactions. An additional problem with 11-dimensional supergravity is that the cosmological equations lead to an anti-deSitter background metric for 4 dimensional spacetime with a very large negative cosmological constant, rather than a Friedman-like universe with a very small cosmological constant. Perhaps even more serious that the cosmological constant problem is the fact that there is no natural explanation for the "Big Bang", i.e. why the observed universe seems to have evolved with time from an initial singularity.

---

*Work performed under the auspices of the U. S. Department of Energy by the Lawrence Livermore National Laboratory under contract No. W-7405-ENG-48.
**Talk presented by George Chapline.

In this talk I will argue that both from the point of view of cosmology and elementary particle phenomenology 10-dimensions seems to be the preferred setting for unification of cosmology and elementary particle physics. On the one hand, 10-dimensional supergravity theories[3] involve a 3-index anti-symmetric field strength, F_{ABC}, which can be used to break the Poincare invariance of the 4-dimensional vacuum and thereby obtain a realistic cosmological model for 4-dimensional spacetime. On the other hand, one can introduce a supersymmetric Yang-Mills theory[4] into 10-dimensions (but not 11-dimensions) and thereby obtain, at least at the tree graph level, a satisfactory description of elementary particle phenomenology. In particular, one can obtain the Glashow-Weinberg-Salam theory of electroweak interactions starting with a supersymmetric Yang-Mills theory in 10-dimensions.[5]

The technical reason that a theory of parity violation is possible in 10-dimensions but not 11-dimensions is that one can have Majorana-Weyl spinors in 10-dimensions. Weyl spinors are, of course, only possible in even dimensions; in addition spinors that satisfy both the Weyl and Majorana conditions are possible in $D = 4n + 2$ dimensions. One can show[5] that this allows flavor chirality in 4-dimensions, even if the 10-dimensional theory is vectorlike. It is particularly intriguing in this respect that a supersymmetric Yang-Mills theory in 10-dimensions necessarily involves Majorana-Weyl spinors,[4] so that parity violation in 4-dimensions is a consequence of supersymmetry in 10-dimensions.

Some other interesting consequences of starting with a supersymmetric E_8 Yang-Mills theory in 10-dimensions are[5-6]: i) SU(5) or SO(10) unification of the strong and electroweak interactions, ii) $\Delta I_W = 1/2$ Higgs field, iii) $5 + 10$ families of fermions, and iv) a natural explanation based on the Atiyah-Singer theorem as to why m_e, m_ν, and $m_d << M_W$.

Given the phenomenological attractiveness of a supersymmetric Yang-Mills theory in 10-dimensions one is naturally tempted to make the theory locally supersymmetric by coupling the $N = 1$ Yang-Mills theory to $N = 1$ supergravity. The Lagrangian for the bosonic fields is this theory has the form[7]:

$$L = \sqrt{-g}\,\left[\frac{1}{2\kappa^2} R - \frac{9}{16\kappa^2}\left(\frac{\nabla\phi}{\phi}\right)^2 - \frac{3}{4}\phi^{-3/2} F_{ABC}^{\;2} - \frac{1}{4}\phi^{-3/4}\,\mathrm{tr}F_{AB}^{\;2}\right] \qquad (1$$

where F_{AB} is the Yang-Mills field strength, ϕ is a scalar field, and κ is the gravity constant in 10-dimensions. An amusing feature of this theory is that the Yang-Mills field is coupled to the antisymmetric tensor field via the Chern-Simons form $X_{ABC} = \mathrm{tr}\{A_{[A}F_{BC]} - 2/3g A_{[A}A_B A_{C]}\}$. This leads to the following identity relating the Yang-Mills field F_{AB} to the supergravity field F_{ABC}:

$$\partial_{[A} F_{BCD]} + \frac{\kappa}{2\sqrt{2}} \mathrm{tr}\{F_{[AB} F_{CD]}\} = 0 \quad . \tag{2}$$

This relation, together with the appearance of the scalar field ϕ in the Yang-Mills part of the lagrangian, shows that 10-dimensions implies a certain "unification" of Yang-Mills type theories and simple supergravity. Unfortunately, this theory suffers from gravitational and non-abelian chiral anamolies, and therefore it is not useful as a quantum theory. On the other hand, one expects that cosmology can be described using classical equations, and thus one might be able to use this theory to obtain a unified description of cosmology and at least some aspects of elementary particle phenomenology.

In order to obtain cosmologically interesting solutions to the D = 10 coupled Yang-Mills supergravity equations one assumes that 10-dimensional spacetime bifurcates into a direct product of the form

$$M^{10} = t \times M^3 \times M^6 \quad ,$$

where M^6 is a compact symmetric space. Following Freund,[1] we will assume that M^3 is distinguished by the appearance of a 3-form field strength

$$F_{ijk} = \varepsilon_{ijk} F \quad . \tag{3}$$

The compact space M^6 is distinguished by the appearance of a Yang-Mills field strength f where

$$f^2 \equiv \mathrm{tr} F^2_{\alpha\beta} \quad . \tag{4}$$

From the point of view of elementary particle physics f^2 may be interpreted as the Higgs potential for the $\Delta I_W = 1/2$ Higgs field appearing in the Weinberg-Salam theory of electroweak interactions.

To obtain the cosmological equations in concrete form one assumes that metric for M^{10} has the form

$$R_{AB} = -\kappa^2 [\tfrac{9}{2}\, e^{-\sqrt{2}\sigma} (F_{ACD} F_B^{\;CD} - \tfrac{1}{12} g_{AB} F^2_{CDE}) \tag{5}$$

$$+ e^{-\sigma/\sqrt{2}}\, \mathrm{tr}(F_{AC} F_B^{\;C} - \tfrac{1}{16} g_{AB} F^2_{CB})\,]$$

$$+ (\nabla_A \sigma)^2 - \tfrac{1}{8} m^2 \sigma^2 \quad ,$$

where we have introduced a new scalar field σ by $\phi = \exp(2/\sqrt{2}\sigma/3)$ in order to reflect the requirement that ϕ should be positive. We have also introduced a mass term for the σ-field; one can show that without such a term there would be no interesting classical solutions. The differential equations for R(t) and S(t) that follow from (5) are given in a recent Physics Letter.[9] Of particular interest is the Raychaudhuri equation

$$3\frac{\ddot{R}}{R} + 6\frac{\ddot{S}}{S} = -\kappa^2\left[\frac{9}{4}e^{-\sqrt{2}\sigma}\frac{F^2}{R^6} + \frac{1}{16}e^{-\sigma/\sqrt{2}}\frac{f^2}{S^4}\right] + \dot{\sigma}^2 - \frac{1}{8}m^2\sigma^2 \quad . \tag{6}$$

By giving the σ-field a mass we have allowed the possibility that $\ddot{R}$ and $\ddot{S} \to 0$ if $S \to S_\infty$ and $\sigma \to \sigma_\infty$; i.e. there is no cosmological constant in the limit where S and σ approach constants. Actually we are only interested in solutions where S and σ approach constants, since only in these cases will phenomenological constants, such as Newton's constant or the W-mass, not vary with time. It is gratifying that not having these constants vary with time is consistent with having a cosmological model for 4-dimensions with no cosmological constant.

As a preliminary to discussing whether there are in fact cosmological solutions to Eq. (5) with these properties let us note that there is a time independent solution $R = R_1$ and $S = S_1$:

$$k_3 = \frac{9\kappa^2}{e^4}\frac{F^2}{R_1^4}$$

$$k_6 = \frac{\kappa^2}{6e^2}\frac{f^2}{S_1^2}$$

$$\sigma = \sqrt{8}$$

$$m^2 = \frac{k_3}{4R_1^2} + \frac{3k_6}{8S_1^2} \tag{7}$$

where k_3 and k_6 are the curvature constants for M^3 and M^6. Evidently time independence requires $k_3 > 0$; to see what happens when $k_3 \leq 0$ we can make use of the Bianchi identity

$$\frac{\dot{R}^2}{R^2} + 5\frac{\dot{S}^2}{S^2} + \frac{\dot{R}\dot{S}}{RS} + \frac{k_3}{2R^2} + \frac{k_6}{S^2} = \frac{3}{2}\kappa^2 e^{-\sqrt{2}\sigma}\frac{F^2}{R^6} + \frac{\kappa^2}{12}e^{-\sigma/\sqrt{2}}\frac{f^2}{S^4} + \sigma^2 m^2 \; . \tag{8}$$

If we assume $S = S_1$ and $\sigma = \sqrt{8}$ the identity becomes

$$\dot{R}^2 - \frac{1}{6}\left(\frac{R_1}{R}\right)^4 - \frac{1}{2}\left(\frac{R}{R_1}\right)^2 = \frac{k_3}{2} \quad .$$

It is clear from the form of the "potential"

$$V(R) = -\frac{1}{6}\left(\frac{R_1}{R}\right)^4 - \frac{1}{2}\left(\frac{R}{R_1}\right)^2$$

that the point $R = R_1$ is an unstable equilibrium point, so that if $\dot{R}$ is positive at $R = R_1$ the universe will expand forever.

In order to investigate possible long term behaviors let us set $R \to \infty$, $S = S_\infty$ and $\sigma = \sigma_\infty$ in Eq. (8):

$$\frac{\dot{R}^2}{R} + \frac{k_6}{S_\infty^2} = \frac{k_6}{2}\frac{S_1^2}{S_\infty^4} + \sigma_\infty^2 m^2$$

and the equation of motion for σ :

$$\sigma_\infty = \frac{3}{\sqrt{8}}\,\frac{1}{(mS_\infty)^2}\left(\frac{S_1}{S_\infty}\right)^4 e^{-\sigma_\infty/\sqrt{2}} \quad .$$

These equations imply that the asymptotic value of the 4-dimensional cosmological constant ${}^{(4)}\Lambda$ is

$${}^{(4)}\Lambda = \frac{m^2\sigma_\infty}{8}\left[\sigma_\infty - \sqrt{8}\right] .$$

One possible solution is $S_\infty = S_1, \sigma = \sqrt{8}$, and ${}^{(4)}\Lambda = 0$.

The physical interpretation of this solution follows from the work of Chapline-Slansky[5] on dimensional reduction of Yang-Mills theories. In particular S_∞ defines a length scale on the order of M_W^{-1}. Also f^2 is determined by the minimum of the Higgs potential for the $\Delta I_W = 1/2$ Higgs field required in the Weinberg-Salam theory of electroweak interactions. We note that $f^2 \neq 0$ is required to obtain $S \to$ constant. It is interesting that $f^2 \neq 0$ is also required to obtain low mass chiral fermions via the Atiyah-Singer theorem.[6] We also note that it is possible to have ${}^{(4)}\Lambda = 0$ even though $f^2 \neq 0$ because of the negative contribution $-k_6/S_1^2$.

We might digress at this point to discuss why we have been able to obtain a solution with no large negative cosmological constant, whereas "spontaneous compactification" of extra dimensions normally leads to anti-deSitter space for

4-dimensional spacetime with a very large negative cosmological constant. The answer can be found by looking at the general form of Raychaudhuri equation

$$R_{00} = T_{00} - \frac{1}{D-2} T^{\sigma}{}_{\sigma} \quad . \tag{9}$$

For a metric of the form (1) we have $R_{00} = {}^{(4)}R_{00}$ where ${}^{(4)}R_{00}$ of the Ricci component for 4-dimensional spacetime. Assuming ${}^{(4)}R_{ij} = {}^{(4)}\Lambda g_{ij}$, i.e., static deSitter or anti-deSitter spacetime, we have

$$-{}^{(4)}\Lambda = T_{00} - \frac{1}{D-2} T^{\sigma}{}_{\sigma} \quad . \tag{10}$$

Thus, if the r.h.s. is positive then ${}^{(4)}\Lambda < 0$. Generalizing the D=4 condition of Hawking and Ellis, we will refer to the positivity of the r.h.s. of (9) as the strong energy condition. The only simple circumstance where the strong energy condition is violated arises when there is a massive scalar field or a scalar field with a potential. Thus the opportunity exists for violating the strong energy condition with any of the D=10 supergravity theories (but not with D=11 supergravity). We took advantage of this opportunity by giving the σ field a mass.

However, it turns out that we must pay a price for a solution without a large negative cosmological constant; namely, our solution with zero cosmological constant is classically unstable. If we write $S = S_0(1+x)$ and $\sigma=\sigma_0(1+z)$ then the small oscillations about a solution where $S=S_0$ and $\sigma=\sigma_0$ are described in the limit $R\to\infty$ by the equations

$$\ddot{x} = m^2\sigma_0\left(\frac{x}{3\sqrt{2}} + \frac{z}{12}\right)$$

$$\ddot{z} = m^2 \left[- 2\sigma_0 x - \left(1 + \frac{\sigma_0}{\sqrt{2}}\right) y\right] \; .$$

It is easy to check that there is always an unstable mode if σ_0 is positive. We believe that this is a general phenomenon, i.e., a flat vacuum for D = 4 is always classical unstable.

Thus in the end our scheme for unifying cosmology and elementary particle phenomenology at the tree graph level has a serious flaw. Specifically, we need a quantum theory of gravity to stabilize the vacuum state. As we have seen 10-dimensions does have a number of phenomenologically attractive features not shared by any other dimension. This leads us to speculate that the fundamental theory may be the theory of closed oriented (Type II) super strings [10]. It is well known that Type II superstring theory has a hidden (non-compact) E_8 symmetry.[10] Less well known is the fact that there is a relation between dual string models and Yang-Mills theories in the limit $D\to\infty$.[11] Therefore it is possible that an E_8 Yang-Mills coupled to supergravity in 10-dimensions might emerge as a classical approximation to Type IIB superstring theory.

Assuming that a D=10 Yang-Mills coupled to supergravity theory does make sense as the classical approximation to a quantum theory and a stable cosmological solution where S and σ approach constants really exists, then the consequences for elementary particle phenomenology that follow from the work of Chapline-Slansky may have some legitimacy. The most spectacular of these predictions is the existence of pyrgons (excitations on M^6) at energies on the order to 1TeV. Thus pyrgons would almost certainly be observable at the SSC, and maybe even at the Tevatron or SPS colliders. A direct kinematic test for the existence of 10-dimensions may also be possible with the SSC. Besides such novel predictions there are also a number of consequences for the standard model of elementary particle physics. If we make some specific choice for the isometry group for the space M^6 then it is possible to make predictions for the $\Delta I_W = 1/2$ Higgs mass and also quark and lepton masses.

In conclusion, we have found that coupling an N=1 Yang-Mills theory to N=1 supergravity in 10-dimensions allows one to obtain a realistic cosmology for 4-dimensional spacetime, and at the same time explain certain aspects of elementary particle phenomenology. However, the existence of a stable cosmological solution with a flat 4-dimensional vacuum requires the existence of a quantum version of the theory. Because of anomalies such a quantum version cannot be obtained using the usual quantization procedures. Instead, one must regard the present theory as an approximation to a more fundamental theory - possibly Type IIB superstring theory.

References

1. P.G.O. Freund, **Nucl. Phys.** B209, 146 (1982).
2. For a reivew see M. J. Duff, **Nucl. Phys.** B219, 389 (1983).
3. A. Chamseddine, **Nucl. Phys.** B185, 403 (1981); M. B. Green and J. H. Schwarz, **Phys. Lett.** 122B, 143 (1983).
4. L. Brink, J. Schwarz, and J. Scherk, **Nucl. Phys.** B121, 77 (1977).
5. G. Chapline and R. Slansky, **Nucl. Phys.** B209, 451 (1982).
6. G. Chapline and B. Grossman, **Phys. Lett.** to be published.
7. G. Chapline and N. Manton, **Phys. Lett.** 120B, 105 (1983).
8. G. Chapline and G. Gibbons, **Phys. Lett.** 135, 43 (1984).
9. J. Schwarz, **Phys. Rep.** 89, 223 (1982).
10. B. Julia, in Proceedings of the May 1982 Erice Conference on Unified Field Theories in more than 4-Dimensions, (Ed.: V. Sabatla, World Scientific 1983).
11. A. M. Polyakov, **Phys. Lett.** 82B, 247 (1979).

DISCLAIMER

This document was prepared as an account of work sponsored by an agency of the United States Government. Neither the United States Government nor the University of California nor any of their employees, makes any warranty, express or implied, or assumes any legal liability or responsibility for the accuracy, completeness, or usefulness of any information, apparatus, product, or process disclosed, or represents that its use would not infringe privately owned rights. Reference herein to any specific commercial products, process, or service by trade name, trademark, manufacturer, or otherwise, does not necessarily constitute or imply its endorsement, recommendation, or favoring by the United States Government or the University of California. The views and opinions of authors expressed herein do not necessarily state or reflect those of the United States Government thereof, and shall not be used for advertising or product endorsement purposes.

REMNANTS FROM COMPACTIFICATION

Edward W. Kolb
Theoretical Astrophysics Group
Fermi National Accelerator Laboratory
Batavia, Illinois 60510
USA

ABSTRACT

Possible observable effects of the existence of compact dimensions are discussed. The effects are remnants from the early stages of the big bang when the Universe was at a temperature comparable to the inverse size of the extra dimensions.

1. Introduction

As discussed in this conference, the origin of the observed internal gauge symmetries from symmetries of an internal compact space is an attractive approach for the unification of particle physics and gravity.[1] The idea of extra dimensions may be implemented by several approaches, a common assumption being that the extra dimensions are compactified to a very small size. Since the natural scale for the extra dimensions is the Planck scale[2], $R_{pl} = 1.6 \times 10^{-33}$ cm, the early Universe may provide the only probe of extra dimensions.

In this paper I will discuss some possible low-energy observable effects of the existence of compactified dimensions. The first effect concerns quantized excitations of the extra dimensions. When viewed from a 4 space-time dimensional world, these excitations correspond to particles with mass comparable to the Planck mass, $m_{pl} = 1.2 \times 10^{19}$ GeV. These excitations should have been produced in the early Universe, and survived annihilation to be present today if they are stable. The second effect is the possibility that extra dimensions are responsible for the observed large entropy of the Universe. Finally, if there exist topological defects in the compactification of the extra dimensions, the defects would appear upon dimensional reduction as massive magnetic monopoles. These monopoles should have been produced in the early Universe and should have survived to the present.

Since the motivation for considering theories with extra dimensions will be reviewed by other speakers at this conference, I will only discuss the above-mentioned low-energy effects.

2. Pyrgons[3]

As pointed out by Klein in 1926[4], the five-dimensional theory of Kaluza implies the existence of an infinite tower of four-dimensional particles corresponding to the non-zero modes of the harmonic expansions in mass eigenstates of the higher-dimensional fields. I will refer to these massive states as PYRGONS, which comes from the Greek πυργος, for tower. Klein's observation was for the 5-dimensional theory, but it is true for all theories based upon dimensional reduction from any number of dimensions.[5] In this talk I will examine the 5-dimensional theory in detail, and then discuss the possible generalization.

In 5 dimensions it is possible to expand all fields in the theory in a harmonic series

$$\Psi_i(x,y) = \sum_{k=-\infty}^{\infty} e^{ik\theta} \phi_i^k (x) \tag{1}$$

where i is a space-time index, and $|k|$ labels the mass eigenstate. The harmonic expansion allows one to separate the dependence upon the 4 non-compact space time dimensions (x^i, $i=0,1,2,3$), and the 5th dimension ($y = 2\pi R\theta$).

The equation of motion for small disturbances about the ground state is[4,5]

$$\pentagon \Psi_i(x,y) = 0 \tag{2}$$

where the "pentagon" operator, $\pentagon$, is $\Box + \partial/\partial y^2$. In the free field limit, each mode ϕ_i^k (x) satisfies the wave equation

$$(\Box + M_k^2)\phi_i^k (x) = 0 \tag{3}$$

where $M_k^2 = (k/2\pi R)^2$. The mass spectrum in four dimensions contains a massless, neutral spin-2 particle; a massless, neutral spin-1 particle; a massless, neutral spin-0 particle; and an infinite tower of charged spin-2 pyrgons with masses $M_k^2 = (k/2\pi R)^2$, $k = 1,2,...$

In the 5-dimensional theory the pyrgons are stable. We label each 4-dimensional field in equation (1) by the quantum number k. The amplitude for the decay process

$$\phi^k \rightarrow \phi^{k_1} + \phi^{k_2} + \ldots + \phi^{k_n} \tag{4}$$

is contained in a term of the 5-dimensional effective action of the form

$$I \sim 2\pi R\int d^4x\int d\theta \, \exp[i(-k+k_1+k_2+....+k_n)\theta] \times \phi^k(x)^* \phi^{k_1}(x)....\phi^{k_n}(x). \tag{5}$$

upon integration over the extra dimension the decay rate for process (4) is proportional to the Kronecher δ, $\delta(k_1 + k_2 + ...k-k)$. This selection rule ensures that the $|k| = 1$ pyrgons cannot decay to the massless $k = 0$ states. However the $k = +1$ pyrgons can annihilate with the $k = -1$ antipyrgons. This selection rule is nothing more than charge conservation. The massive ($m \simeq m_{pl}$) states are charged while the massless ($m \ll m_{pl}$) states are neutral.

If the Universe was ever at temperatures comparable to R^{-1} (R is the compactification scale) the pyrgons should have been present.

If the Universe was ever at temperatures in excess of R^{-1}, the Universe would have been effectively 5-dimensional, and since all the excitations of the geometry are "massless" in 5 dimensions [cf. equation (2)] they will be distributed over all the modes in equation (1). If the maximum temperature of the Universe was less than but comparable to R^{-1}, the pyrgon states should have been pair produced. In either case, a reasonable guess for initial conditions might be the first few modes in the expansion as abundant as the zero modes, which will be denoted as photons.

The excited pyrgons ($|k|>1$) decay rapidly into $|k| = 1$ pyrgons and $k = 0$ photons. The $|k| = 1$ pyrgons cannot decay, and their co-moving number density is decreased only by annihilation. Slansky and I have shown that annihilation is ineffective in riding the Universe of pyrgons and there should be as many pyrgons as photons in the Universe[3]. If there are Planck-mass pyrgons in the Universe as abundant as photons, the mass density of the pyrgons would be $\rho \simeq 8 \times 10^{-3}$ g cm^{-3}, or about 4×10^{26} times the critical density. Obviously some mechanism must either suppress pyrgon production, rid the Universe of pyrgons, or create a lot of entropy so as to dilute the relative pyrgon density.

The most likely mechanism to rid the Universe of pyrgons is decay. The reason they were stable in the 5-dimensional example is that they carried a charge not carried by the zero modes. However, there are somewhat more realistic models, such as 11-dimensional supergravity with vacuum $M^4 \times S^7$,[6] that have stable pyrgons even though most of the 256 zero modes do carry the SO_8 quantum numbers of the symmetry of S^7.[3] It is possible to imagine several reasons for assuming some pyrgons might be stable. For example if all zero modes satisfy the usual electric charge/color relation, but some pyrgon does not. One might also imagine that there is an additional charge under which all the zero modes are neutral, but some pyrgon is not. In general, until a successful model is constructed and the mass spectrum computed, it is not possible to know whether there will be stable pyrgons.

Although it is not possible to know the properties of the stable pyrgons (if they exist) without the benefit of a model, a good guess might be that they would appear as massive, colorless, fractionally-charged particles. Such particles would be very massive $m \sim 10^{17} - 10^{19}$ GeV, and today might appear as penetrating particles with velocities comparable to the galactic virial velocity ($v \approx 10^{-3}$ c). It may be possible to detect such particles in ionization monopole searches, since at the expected low velocities they would have an energy loss similar to magnetic monopoles.[7]

3. The Entropy of the Universe[8]

The origin of the observed large entropy of the Universe is the outstanding problem in cosmology today. If it is possible to create a large amount of entropy ($S \gtrsim 10^{86}$) in a causal (hence possibly smooth) region, it may offer an explanation of the observed homogeneity/isotropy, oldness/flatness, and horizon problems.[9]

There have been two approaches to exploit extra dimensions to produce entropy. The first approach involves the Universe going through a de-Sitter phase caused by the effective vacuum energy necessary for compactification.[10] This approach depends upon a non-adiabatic era in which vacuum energy is converted into thermal energy.

In this talk I will discuss a second approach based upon the conversion of the entropy in the extra dimensions to entropy in the three observed dimensions.[8,11,12] Such an approach results in entropy production that is "adiabatic" in the sense that it is assumed that the entropy in the total co-moving spatial volume is constant.

To illustrate this effect, consider a toy model with one time and 3 + D spatial dimensions and a metric with symmetry $R^1 \times S^3 \times S^D$:

$$g_{MN} = \mathrm{diag}\,(1,\ R_3^2\tilde{g}_{mn},\ R_D^2\tilde{g}_{\mu\nu}) \tag{6}$$

where R_3 and R_D are the cosmological scale factors for S^3 and S^D, and $\tilde{g}_{mn}$ and $\tilde{g}_{\mu\nu}$ are metrics for maximally symmetric 3-dimensional and D-dimensional spaces. For simplicity we will assume $\tilde{g}_{mn}$ is flat, i.e. $S^3 \to R^3$. If a large amount of entropy can be converted from S^D to S^3, the limit $S^3 \to R^3$ would be reasonable as entropy creation leads to a "flat" Universe. We will assume the stress-energy tensor takes the 3 + D-dimensional perfect fluid form, and is <u>isotropic</u>,[13] i.e. it can be described by an energy density, ρ, and a pressure, p:

$$T_{MN} = p g_{MN} + (p + \rho)U_M U_N \quad . \tag{7}$$

With the above metric and stress-energy tensor, the dynamical equations of motion for the two scale parameters are given by ($N = 3+D$)

$$3\ \ddot{R}_3/R_3 + D\ \ddot{R}_D/R_D = 8\pi G[(2-N)\rho - pN]/(N-1) \tag{8a}$$

$$d/dt(\dot{R}_3/R_3) + \dot{R}_3/R_3\ (3\ \dot{R}_3/R_3 + D\ \dot{R}_D/R_D) = 8\pi\bar{G}(\rho-p)/(N-1) \tag{8b}$$

$$R_D^{-2} + d(\dot{R}_D/R_D)/dt + \dot{R}_D/R_D(3\ \dot{R}_3/R_3 + D\ \dot{R}_D/R_D) = 8\pi\bar{G}\ (\rho-p)/(N-1) \tag{8c}$$

where $\bar{G}$ is related to Newton's constant, G_N, by $\bar{G} = G_N V_D$ with V_D the volume of the compact space today.

Since in the extra dimensional picture all fields are massless, we will assume the equation of state for radiation, $p = \rho/N$. The assumption of an isotropic equation of state may fail for several reasons. One reason for failure would be if the difference in expansion rate between R_3 and R_D is larger than the rate for equilibration of the particles. If $|\dot{R}_3/R_3 - \dot{R}_D/R_D| > \Gamma$, where Γ is a typical interaction rate, the energy will not be able to be distributed along different directions, and the perfect fluid approximations would fail. Proper consideration of the departure from the perfect fluid behavior would involve complicated transport calculations. Instead, I will assume perfect fluid behavior until a critical point is reached when the wavelength of the excitations of the compact dimensions ($\lambda \propto T^{-1}$) is equal to the radius of the compact space, R_D. I will assume that freeze-out of the extra dimensions occurs when $R_D T = 1$.

The program to calculate the entropy is straightforward. The field equations have a 1-parameter family of solutions. For each solution we integrate the equations of motion until $R_D T = 1$, then calculate the entropy in the horizon 3-volume. The horizon distance, ℓ_H, for S^3 is given by

$$\ell_H(t) = R_3 \int_0^t dt'/R_3(t') \quad . \tag{9}$$

The energy density in N dimensions is $\rho_N = T^{N+1}$. At freeze out, all the degrees of freedom (entropy) in the extra dimensions is converted to degrees of freedom in S^3. (Barr and Brown point out that this effect is nothing more than the decay of the pyrgon states.) If we denote the value of R_D when $R_D T = 1$ by R_*, then the effective 3-dimensional energy density at freeze-out is $\rho_3 = R_*^D \rho_N = R_*^{-4}$, since at freeze-out $T = R_*^{-1}$. The 3-dimensional entropy density is given by $s_3 = \rho_3^{3/4} \simeq R_*^{-3}$, and the total 3-entropy is $S_3 = (\ell_H R_D^{-1})^3$.

The generic behavior of the two scale factors R_3 and R_D are shown in Figure 1. The Einstein equations do not admit a solution with static compact dimensions. Therefore the compact dimensions reach a maximum then decrease, eventually approaching a singularity. As R_D approaches the final singularity, R_3 diverges. Because of the collapsing extra dimensions the mean volume ($V^N \propto R_3^3 R_D^D$) decreases, which leads to an increase in temperature since in an adiabatic expansion (or collapse) $T^N V^N$ is constant. The increase in temperature while R_3 increases leas to an increase in the entropy temperature while R_3 increases leads to an increase in the entropy in the 3-horizon volume.

In Figure 2 we show the entropy in the 3-horizon volume when $R_D T = 1$, for different numbers of compact dimensions. There is a one-parameter family of solutions for a given number of dimensions. We parameterize this family of solutions by R_*, the physical radius of the extra dimensions when $R_D T = 1$. We would expect the freeze-out radius to be comparable to the radius of the extra dimensions today, hence we expect $R_* \gtrsim R_{pl}$. For 2 extra dimensions the asymptotic form of S_3 is $S_3 \sim (R_*/R_{pl})^{27}$, so if $R_* \gtrsim 3000\ R_{pl}$, there will be sufficient entropy production. For more than 2 extra dimensions, the slope becomes negative, e.g. for 3 extra dimensions $S_3 \sim (R_*/R_{pl})^{-50}$, which suggests that R_* must be less than R_{pl} for sufficient entropy production. As the number of dimensions increases the slope becomes more negative.

We believe that there are more than two extra dimensions and that the radius of the extra dimensions should be greater than R_{pl}, so our toy model failed. However the basic mechanism works -- entropy does increase in the horizon 3-volume. Perhaps adding external fields or a cosmological constant to stabilize extra dimensions would change the equations of motion enough to allow sufficient entropy production. Abbott, Barr, and Ellis[12] have pointed out that if we relax the freeze-out condition, it is possible to create more entropy. In fact they point out that if you go to a large number of extra dimensions the entropy contained within a volume of radius R_D increases with increasing number of extra dimensions, $S \sim (R_D T)^D$, and if D is large, $D > 44$, then for $R_D T$ reasonable, say $R_D T > 100$, S would be greater than 10^{88}. In other words for a large number of dimensions a huge volume (hence entropy) can be characterized by a small radius.

4. Kaluza-Klein Monopoles[14]

Gross and Perry[15], and also Sorkin[16] have shown that topological defects in the geometry of compactification would appear in four space-time dimensions as massive magnetic monopoles.

GUT monopoles are produced in the early Universe when topological defects are frozen in during a phase transition associated with spontaneous symmetry breaking. Kaluza-Klein monopoles are produced when topological defects are frozen in as space is split into 3 large spatial dimensions, and D small, compact dimensions. There may be a fundamental difference between GUT monopole production and Kaluza-Klein monopole production. In the standard cosmology the Universe was in the symmetric phase as the initial condition and there were no GUT monopoles. The monopoles first appear during the phase transition, so the number of monopoles produced is independent of any monopole initial conditions. However if the splitting of the spatial dimensions is present as an initial condition, then Kaluza-Klein monopoles appear as initial conditions, and it is impossible to calculate the expected number of monopoles. Therefore in cosmological models where the splitting is an initial condition, we cannot predict the expected monopole abundance. However if the splitting is dynamical, then the number density of monopoles can be estimated.

To illustrate the calculation of monopole production if compactification is dynamic, let us assume that the Universe expands isotropically in $N = D + 3$ spatial dimensions from $t = 0$, until $t = t$ when there is a fluctuation in the geometry that causes D dimensions to compactify. In this case one may regard compactification as a phase transition and use the horizon distance as the order parameter. This would result in an average production of one monopole per horizon volume. In an isotropic N-dimensional, radiation-dominated expansion, the horizon distance at compactification is

$$\ell_H(t) = t = \frac{N+1}{2}\left[\frac{N(N-1)\Gamma(N/2)\Gamma(1+D/2)}{32\pi^{2+N+D/2}\Gamma(N+1)}\right]^{\frac{1}{2}} \frac{m_{Pl}\, m_{KK}^{D/2}}{T} \tag{10}$$

where t is the time from the initial singularity, and we have assumed an isotropic expansion in $N = 3 + D$ spatial dimensions with a radiation dominated equation of state $\rho = [2\pi^{N/2}/\Gamma(N/2)]\ T^{N+1} = Np$. We assume that compactification occurs at $T = m_{KK}$, so that

$$\ell_H(t_c) \simeq m_{pl}/m_{KK}^2 \tag{11}$$

which results in a monopole density of

$$n_M(T_c) = \ell_H^{-3}(T_c) \simeq m_{KK}^6/m_{pl}^3 \quad . \tag{12}$$

If we assume that compactification is instantaneous the photon density at compactification ($T = R_{KK}^{-1}$) would be

$$n_\gamma = V_D \rho_N \simeq R_{KK}^D\ T^{N+1}$$
$$= m_{KK}^3 \qquad\qquad (R_{KK} = T). \tag{13}$$

Therefore the monopole to photon ratio in this model would be given by

$$n_m/n_\gamma = (m_{KK}/m_{pl})^3 \quad , \tag{14}$$

which is the same as the prediction for GUT monopoles if one takes $m_{KK} = m_{GUT}$.

5. Conclusion

Finding a "Kaluza-Klein" approach which gives sensible low-energy physics is presently a problem being considered by many people. If tomorrow someone discovers a theory that gives the correct observed low-energy physics (including fermions!) have they proved the existence of extra dimensions? What is needed is some test for the existence of extra dimensions that is accessible via low-energy physics. In this paper I have described three possible effects of the existence of extra dimensions that are in some sense "low-energy" effects. These low-energy effects are present today because the Universe was once at high energy. The discovery of pyrgons or monopoles with mass of the order of the Planck mass would be strong evidence for the existence of extra dimensions. Perhaps the observed large entropy of the Universe is due to the existence of extra dimensions.

6. References

1. Th. Kaluza, Sitzungsber. Preuss. Akad. Wiss. Berlin, Math. Phys. KI(1921)966; O. Klein, Z. Phys. 37(1926)895.

2. Theories with the scale of the extra dimensions comparable to $m_W \simeq$ 100 GeV have also been proposed, see e.g. G. Chapline and R. Slansky, Nucl. Phys. B209, 461 (1982).

3. E. W. Kolb and R. Slansky, Phys. Lett. 135B, 378 (1984).

4. O. Klein, Nature 118, 516 (1926).

5. A. Salam and J. Strathdee, Ann. Phys. (NY) 141, 316 (1982).

6. R. D'Auria and P. Fre, Phys. Lett. 121B, 141 (1983); M. Duff, B. E. W. Nilsson and C. N. Pope, Phys. Rev. Lett. 50, 2043 (1983).

7. See e.g. N. Nashimo, et. al. Phys. Lett. 128B, 327 (1983).

8. E. W. Kolb, D. Lindley and D. Seckel, Fermilab preprint Pub-84/37-AST.

9. For a review, see M. Turner "Inflation Circa 1983", in Proceedings of IV Workshop on Grand Unification, eds. H. A. Weldon, P. Langacker, and P. J. Steinhardt (Birkhauser, Boston, 1983) p. 228.

10. Q. Shafi and C. Wetterich, Phys. Lett. 129B, 387 (1983).

11. E. Alvarez and M. Belen-Gavela, Phys. Rev. Lett. 51 (1983)931; M. Böhm, J. L. Lucio, and A. Rosado, "Dynamical Compactification an Alternative to Inflation?" Preprint (1983); S. Barr and L. Brown, "On Entropy from Extra Dimensions", University of Washington Report 40048-01 P4 (1983); D. Sahdev, Phys. Lett. 137B (1984) 155.

12. R. B. Abbott, S. M. Barr, and S. D. Ellis, University of Washington preprint 40048-03 P4 (1984).

13. For the limitations of this assumption, see M. Dresden, et. al., these proceedings.

14. J. A. Harvey, E. W. Kolb, M. Perry, Princeton University preprint (1984).

15. D. Gross and M. Perry, Nucl. Phys. B226, 29 (1983).

16. R. Sorkin, Phys. Rev. Lett. 51, 87 (1983).

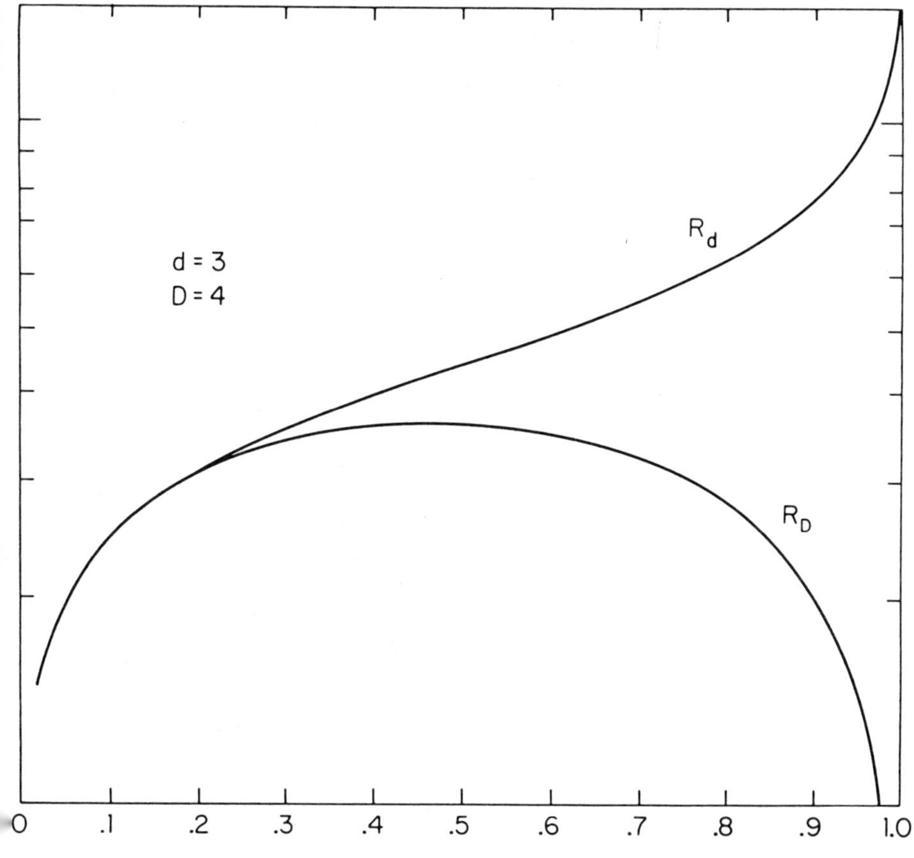

Figure 1

The compact (R_D) and open (R_d) scale factors as a function of x where x is the dimensionless variable proportional to the time, and x_s is the position of the final singularity. Note that both go to zero in the same way at x = 0. At $x = x_s$, $R_D \to 0$ while $R_3 \to \infty$.

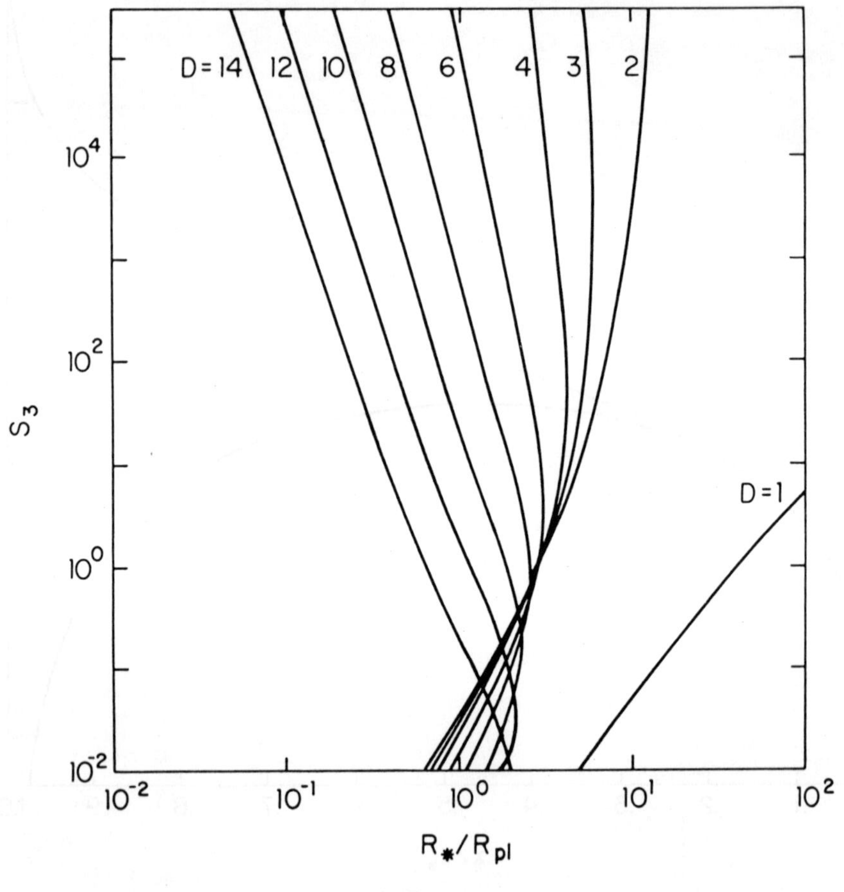

Figure 2

The entropy in the horizon volume of the 3 non-compact dimensions, S_3, as a function of R_*/R_{pl}.

TOWARDS SUCCESSFUL PHASE TRANSITIONS IN THE EARLY UNIVERSE

Jai Sam Kim

Department of Physics & Astronomy
The Johns Hopkins University
Baltimore, MD 21218

ABSTRACT

We suggest a way out of the phase transition problem observed by Breit, Gupta and Zaks and Moss in the new inflationary universe theory. As a by-product, we find a severe constraint on the choice of scalar representations in grand unification theories that employ the Coleman-Weinberg symmetry breaking mechanism or the Higgs mechanism with cubic terms included.

The grand unification theories[1)] were found to give a natural explanation[2,3)] to some outstanding puzzles in the standard big bang cosmology: the horizon puzzle, the flatness puzzle, the baryon excess puzzle. The Friedman universe cools into the symmetry preserving vacuum. As the temperature falls, the symmetric vacuum becomes unstable and the symmetry breaking vacuum becomes stable. The phase transition from the false to the true vacuum is not instantaneous. During the course of transition, there exists a short period when the expansion rate of the scale factor stays virtually constant. The universe enters into the de Sitter phase and expands exponentially, solving the horizon puzzle.

If the vacuum evolves in the SU(3) x SU(2) x U(1) phase during the inflation, the monopoles that might be abundantly created[4)] are diluted, solving the monopole problem. In the early works[2,3)] on the inflationary universe, it was assumed that the vacuum evolved along the SU(3) x SU(2) x U(1) direction from the symmetric zero point to the absolute minimum of the potential.

Last year it was observed[5)] that the Coleman-Weinberg potential with the SU(5) adjoint of scalars does not allow the universe to inflate in the SU(3) x SU(2) x U(1) phase. Later it was shown[6] that the Coleman-Weinberg potential with the scalars assigned to the adjoint representations of SU(N) (N>3) and SO(2N) ($N\neq 4,8$) drive the universe into the undesired phases, SU(N-1) x U(1) and SO(2N-2) x U(1) respectively. It was suggested[5,6,7] that this disease might be cured if the scalar self-coupling constant is made large, which invalidates the one loop approximation.

The problem of evolving into the wrong vacuum is wide-spread in many unification models using nontrivial gauge multiplets. As we will show later in this paper, it occurs in the classical Higgs potential with cubic terms included, which might be unavoidable in the Kaluza-Klein inflation models[8]. The GUT phase transition is a nontrivial problem in supersymmetric inflation models.[9] It was suggested[10] that the "confinement" of SU(5) interactions shifts the height of the effective potential by a different amount, $-[(\pi^2/90)(N_B + (7/8)N_F)]T^4$, in each different phase. This argument implies that the effective potential sharply (like a δ-function) drops in the SU(3) x SU(2) x SU(1) direction out of the SU(3) x U(1) x U(1) or SU(2) x SU(2) x U(1) x U(1) directions, which is rather unnatural.

The disease can be tuned away if the classical Higgs potential without cubic terms is used. There is no direction problem if the gauge singlet[11] or other trivial multiplets with only one little group are used. It was shown[12] that additional slow rollover inflation can occur during the phase transition from an undesired intermediate phase to the desired stable phase. In order to realize this mechanism, one needs to drive the universe into a false vacuum (not into the SU(3) x SU(2) x U(1) phase, otherwise barely achieved baryon excess can be wiped out in the ensuing inflation). Since it takes at least two irreducible representations and the maximal little group of both representations is normally smaller than SU(3) x SU(2) x U(1), the mechanism is not likely to work for SU(5). It may be a viable mechanism for large gauge groups.

It was shown[6] that the Coleman-Weinberg potential for an irreducible representation has the lowest profile near the zero point in the direction where $\alpha \equiv \mathrm{Tr}M^4/(\|\phi\|)^2$ is maximized. (M is the gauge boson matrix.) For the SU(5) adjoint, this direction happens to be SU(4) x U(1). A direct phase transition from the SU(4) x U(1) to the SU(3) x SU(2) x U(1) phase is not possible because these phases are isolated in the sea of possible phases. The vacuum has to go through intermediate phases until it arrives at the SU(3) x SU(2) x U(1) phase. Naturally monopoles are created. By this time, the universe is already out of inflation and the monopoles cannot be diluted. However, if α is maximized in the SU(3) x SU(2) x U(1) direction for some other choice of the scalar representation, the vacuum can be made to evolve along the SU(3) x SU(2) x U(1) direction from the zero point to the absolute minimum with a reasonable choice of scalar self-coupling constants.

The requirement that α be maximized in the SU(3) x SU(2) x U(1) or other desired direction restricts the choice of scalar representations fairly severely. The above requirement is only necessary. The sufficient condition for a successful phase transition will be given later in this paper.

In order to find out the direction the vacuum begins to roll down, let us take a look at the evolution equations. Solving the evolution equations with many scalar fields is highly non-trivial. However, a small piece of knowledge about the potential structure can reduce it to a small dimensional problem in the inflation region. The Coleman-Weinberg potential for scalars, ϕ, χ, etc., assigned to various irreducible representations can generally be written in the following simple form (in units of some mass scale μ);

$$V_{c-w} = \alpha R^4 \ell n R^2 + \beta R^4 \tag{1}$$

where $R^2 \equiv \|\phi\| + \|\chi\| + \ldots$, $\|\phi\| \equiv X_\phi R^2$, $\|\chi\| = X_\chi R^2$, ..., $X_\phi + X_\chi + \ldots =$, and (α,β) are functions of $(\hat{\phi}_i, \hat{\chi}_j, \ldots; X_\phi, X_\chi, \ldots)$ with $\hat{\phi}_i \equiv$ $_i/\|\phi\|^{1/2}$, $\hat{\chi}_j \equiv \chi_j/\|\chi\|^{1/2}$, etc. The evolution equations of the scalar vacuum expectation values are, neglecting radiation damping;

$$\ddot{\phi}_i + 3H\dot{\phi}_i + \frac{\partial V}{\partial \phi_i} = 0, \quad \ddot{\chi}_j + 3H\dot{\chi}_j + \frac{\partial V}{\partial \chi_j} = 0 \text{ , } \ldots \tag{2}$$

Since the equations (2) for ϕ_i and χ_j are formally identical, they can be put in a single form,

$$\ddot{\Phi}_i + 3H\dot{\Phi}_i + \frac{\partial V}{\partial \Phi_i} = 0 \tag{2'}$$

with $\Phi_i = \phi_i$, $\Phi_{n+j} = \chi_j$, ... The reason we have to treat the potential part separately is that ϕ and χ do not form a single vector in the representation space. The Einstein equation,

$$H^2 = (8\pi/3m_p^2)\ [V(\Phi) + \frac{1}{2} \sum_i \dot{\Phi}_i^{\ 2}] \text{ ,} \tag{3}$$

becomes non-trivial in the post inflation period.

During the inflation when the "slow rollover" conditions [13], $\ddot{\Phi}_i$ 0, and $H \simeq$ constant, are satisfied, the vacuum rolls down along the negative gradient direction in the Φ_i-space. The Coleman-Weinberg potential satisfies the slow rollover condition in the region where $R \leqslant 10^{-4}$. The profile of the Coleman-Weinberg potential in this region can be analyzed fully using the technique given in ref. (6). At a given R, the lowest value of V_{C-W} is found at the point where the line $\beta = -(\ell nR^2)\alpha + k_\ell/R$ makes first contact with the "orbit space" (α,β) as k_ℓ is increased from $-\infty$.

If the point with α_{max} is a <u>cusp</u> in the orbit space boundary, the potential is the lowest along the direction of α_{max} in the region $0 \leqslant R \leqslant R_t$ where $R_t \gg 10^{-4}$ for a wide range of scalar coupling constants. At some R_c, the line can meet other points with non-maximal α first. However the point of α_{max} remains locally lowest until the slope of the line at R_t becomes equal to the tangent $d\beta/d\alpha$ of the lower boundary curve at that point. For an irreducible representation, the evolution equation is one dimensional in the region $0 \leqslant R < R_t$. For a reducible representation, R normally depends on $(X_\phi, X_\chi, \ldots)$. If $V_{C-W}(R, X_\phi, X_\chi, \ldots)$ at $R < R_t$ is minimized with all X's non-zero, the evolution equation contains one component from each representation (ϕ, χ, ...). In the case of the SU(5) 24 + 5 + $\bar{5}$ [14], we have found that V_{C-W} is minimized with the $\bar{5}$+5 scalars zero in the inflation region.

Whether the point of α_{max} is a cusp is determined by group theory only, regardless of the scalar coupling constants. If the representation is irreducible, α is normally maximized in the direction(s) of maximal little group(s)[15,16,17], where β is also extremized. Thus the orbit space near α_{max} is wedge shaped. If α_{max} is realized in the directions of several maximal little groups[18], then the lowest profile is in the direction of the maximal little group with smaller β. For a reducible representation the orbit space structure is not fully understood. In this case there are less cusps. For SU(N) adjoint + vector, there is one direction, namely SU(N-1), where α is maximized and β is extremized. In this case, the orbit space near this point is again wedge shaped.

In order to obtain the correct phase transition, one needs to adjust the coupling constants such that the direction of α_{max} stays lowest well beyond the absolute minimum ($R_t > 1$) which should necessarily be in this direction. Thus if we find a good representation in which the cusp of α_{max} is along the SU(3) x SU(2) x U(1) direction and adjust the coupling constants to make the absolute minimum occur in this direction, then the evolution equation contains the fewest scalar components and the phase transition problem is solved.

If the point of α_{max} is <u>not a cusp</u>, the line will sweep the orbit space boundary curve as R increases and the direction minimizing V_{C-W} changes continuously. The evolution equation contains at least two scalar components. Its solution will depend on the scalar coupling constants sensitively. As long as the point of α_{max}, the absolute minimum and the boundary curve joining them preserve the same little group (up to conjugation), the monopole problem can be avoided. We believe this is not the case in reality.

If we were to use the Coleman-Weinberg mechanisim, the above requirement can almost uniquely single out the scalar representation. There are three representations of dimension less than 1000 in SU(5), 24(1001), 75(0110) and 200(2002), allowing SU(3) x SU(2) x U(1) breaking. 24 is already excluded. 200 is also excluded for the following reason. For our purpose it sufficies to know ratios of α along each direction. Since TrM^2 is a group invariant quadratic polynomial, it must be proportional to $\|\phi\|$. Thus $\alpha = c\, TrM^4/(TrM^2)^2$. 200 has three maximal little groups; SU(3) x SU(2) x U(1) with 24 = (1,1)(0) + (1,3)(0) + (8,1)(0) + (3,2)(-5) + ($\bar{3}$,2)(5), SU(4) x U(1) with 24 = 1(0) + 15(0) + 4(-5) + $\bar{4}$(5), and SO(5) with 24 = 10 + 14. The diagonalized gauge boson $(mass)^2$ matrix has 12, 8, and 14 identical elements along the SU(3) x SU(2) x U(1), SU(4) x U(1) and SO(5) directions respectively. The ratio of α's is 1/12:1/8:1/14. Thus 200 favors SU(4) x U(1). This simple trick is not usable for 75 because the elements of the M^2 matrix are not all identical along some directions, e.g. Sp(4) x U(1). Detailed analysis will be reported elsewhere[19].

The classical Higgs potential without the cubic term can be made (by adjusting coupling constants) to have the desired shape. The vacuum rolls down along the SU(3) x SU(2) x U(1) direction and then to the SU(3) x U(1) minimum. However, the cubic term (if the representation allows it) can be troublesome. (It is not generated by radiative corrections nor by finite temperature corrections and thus not indispensible in grand unification theories.) Let us write the potential in the form: $V_{c\ell} = -m^2R^2 + \beta R^3 + \alpha R^4$. Near the zero point, the R^4 term is negligible unless β is extremely small compared to α. The potential is the lowest in the direction of $|\beta|_{max}$ near the zero point. This direction can be an undesired one. In the SU(5) adjoint case, it is SU(4) x U(1), in the 75 case, Sp(4) x U(1) and in the adjoint + vector case, SU(4). To avoid the difficulty, we need to find a good representation where $|\beta|_{max}$ is realized at SU(3) x SU(2) x U(1) or SU(3) x U(1).

I am grateful to Professor C.W. Kim and Mr. P. Murphy for helpful discussions and Professors G. Feldman and L. Madansky for encouragement. This work was supported in part by the National Science Foundation.

References

1) J.C. Pati and A. Salam, Phys. Rev. Lett. 31 (1973), 661; H. Georgi and S. Glashow, Phys. Rev. Lett. 32 (1974), 438.

2) A.H. Guth, Phys. Rev. D23 (1981), 347.

3) A.D. Linde, Phys. Lett. 108B (1982), 389; A. Albrecht and P.J. Steinhardt, Phys. Rev. Lett. 48 (1982), 1220; S.W. Hawking and I.G. Moss, Phys. Lett. 110B (1982), 35.

4) J.P. Preskill, Phys. Rev. Lett. 43 (1979), 1365.

5) J.D. Breit, S. Gupta and A. Zaks, Phys. Rev. Lett. 51 (1983), 1007; I.G. Moss, Phys. Lett. 128B (1983), 385.

6) J.S. Kim and C.W. Kim, Nucl. Phys. B (1984), to appear.

7) M. Sher, Phys. Lett. 135B (1984), 52.

8) M. Bowick, private communication (1984).

9) D.V. Nanopoulos, K.A. Olive, M. Srednicki, and K. Tamvakis, Phys. Lett. 123B (1983), 41;
A. Albrecht, S. Dimopoulos, W. Fischler, E.W. Kolb, S. Raby and P.J. Steinhardt, Nucl. Phys. B229 (1984), 528;
B.A. Ovrut and P.J. Steinhardt, Phys. Lett. 133B (1983), 161.
R. Holman, P. Ramond, and G.G. Ross, Phys. Lett. 137B (1984), 343.
G. Gelmini, C. Kounnas, and D.V. Nanopoulos, Primordal Inflation with Flat Supergravity Potentials, CERN-TH3777, (Dec. 1983).

10) D.V. Nanopoulos, K.A. Olive, and K. Tamvakis, Phys. Lett. 115B (1982), 15.

11) Q. Shafi and A. Vilenkin, Phys. Rev. Lett. 52 (1984) 691; S-Y. Pi, Phys. Rev. Lett. 52 (1984), 1725;
P.Q. Hung, Cascading Inflationary Universe..., U. Virginia preprint (Nov. 1983).

12) S. Gupta and H.R. Quinn, Phys. Rev. D (1984), to appear.

13) P.J. Steinhardt and M.S. Turner, Phys. Rev. 29 (1984), 2162.

14) K. Kang, C.W. Kim, and J.S. Kim, to be published.

15) L. Michel, C.R. Acad. Sc. Paris 272 (1971), 433; CERN-TH 2716 (1979).

16) M. Abud and G. Sartori, Ann. Phys. 150 (1983), 307.

17) J.S. Kim, J. Math. Phys. 25 (1984), 1694.

18) C.J. Cummins and R.C. King, Symmetry Breaking Patterns..., Southampton preprint (1984).

19) C.W. Kim, J.E. Kim and J.S. Kim, to be published.

SUPERSYMMETRIC INFLATIONARY COSMOLOGY

P. Ramond

Physics Department, University of Florida,
Gainesville, Florida 32611

The standard cosmological model[1] for the formation of the elements has proven increasingly successful with the explosion of astrophysical data over the last two decades. According to it, gravity is described by the homogeneous, isotopic metric, of Friedman, Robertson and Walker, it contains one time dependent scale parameter R(t) and can occur in three varieties corresponding to an open, closed or critical expansion of the universe. Einstein's equation then demands that matter be described as a perfect cosmic fluid with energy density $\rho(t)$ and pressure p(t). There is no reason to doubt the adequacy of this picture from the time of cosmological nucleosynthesis to the present, provided one includes density fluctuation $\delta\rho$ to account for the observed inhomogeneities we are a part of. At that time, the assumption that the matter part was dominated by a heat bath at a temperature T of the order of Mev (corresponding to the p-n mass difference and inversely proportional to R) seems to be successful. The question of interest is to determine what happened before that time, and in order to answer this question one needs information as to the type of matter and its behavior at those early times; for this, one turns to particle physics. In fact the second marriage of particle physics and cosmology was due to the observation[2] that Grand Unification provided a ready-made mechanism for generating baryon number asymmetry. Armed with a specific model of particle physics at high energy, one can start exploring the earlier epochs of the universe. No model with sufficiently high credibility exists but we will assume that it incorporates supersymmetry as a way of preserving very small scale ratios in the presence of radiative corrections. We call such a model Supersymmetric Quantum Interacting Dynamics (SQID). Before discussing specific SQIDs let us recall some elementary facts about the FRW description of cosmology.

Einstein's equations reduce to

$$\left(\frac{\dot{R}}{R}\right)^2 + \frac{K}{R^2} = \frac{1}{3M^2}\,\rho(t), \tag{1}$$

where $M = 2.4 \times 10^{18}$ GeV, and we define it to be one Planck (P), and $K = 0,\pm1$ depending on the eventual fate of R(t). The matter density $\rho(t)$ is govered by

$$\dot{\rho}(t) + 3\left(\frac{\dot{R}}{R}\right)(\rho+p) = 0, \tag{2}$$

where p is the pressure. These equations then have to be augmented by giving the equation of state of matter, i.e. by expressing ρ as a function of p. We summarize some interesting cases:

a) Relativistic matter $\rho = \frac{1}{3} p$ gives $\rho \sim R^{-4}$;

$H \equiv \left(\frac{\dot{R}}{R}\right) \sim t^{-1}$; $R(t) \sim t^{1/2}$.

b) Non Relativistic matter $\rho >> p$ gives $\rho \sim R^{-3}$;

$H \sim t^{-1}$; $R(t) \sim t^{2/3}$.

c) Cosmological matter $\rho = -p =$ constant;
it gives $H \approx \frac{1}{3M^2} \rho_0$ $R(t) \sim e^{Ht}$.

One can envisage other interesting cases, such as "stiff matter" for which $\rho = p$, but popular particle physics scenarios seem to produce only cases a), b) and c).

Although matter at least from the time of cosmological nucleosynthesis is described by thermal considerations, it must be emphasized that this is so only because at those times the interaction rate among particle species is in general much faster than the expansion rate of the universe, thus allowing for the particles to share their energy and for the notion of temperature to emerge. Thus one has to continuously check whether

$$H \lessgtr \langle n\sigma v \rangle \tag{3}$$

is satisfied or not, where σ is the cross section, n the density and v the velocity. We will see that it is not inconceivable to have $\rho(t)$ at some earlier time dominated by particles which do not have time to achieve equilibrium due to the weakness of their interactions.

Having assembled these tools, we can now start to describe the demands on the right-hand side of Einstein's equation set both by cosmology and by particle physics.

At the time of, say, cosmological nucleosynthesis, various parameters must be given initial values in order to successfully describe the universe today. In particular the observed isotropy and homogeneity of the detected microwave background and the observed value of the mass density suggests that there was an enormous amount of entropy[3] in the very early universe. The problem is how to explain its origin by using a credible particle physics model. Such a large initial entropy would help explain why the observed universe is so close to criticality and yet so old; namely, if you define the critical density, ρ_c, via Hubble's constant as

$$H^2 = \frac{1}{3M^2} \rho_c, \tag{4}$$

the observed luminous density ρ_ℓ is seen to be

$$\Omega_\ell \equiv \frac{\rho_\ell}{\rho_c} \sim 10^{-2},$$

while dark matter around which luminous matter congregates is estimated to be

$$\Omega_d \equiv \frac{\rho_d}{\rho_c} \sim 2 \times 10^{-1}.$$

Now this value is very close to 1 (in cosmology, errors are usually quoted in the exponent), and the evolution equation (1) shows that given the "age of the universe", Ω would have been to be equal to 1 in one part in 60(!) given adiabaticity. However a severe loss of adiabaticity, such as a large entropy release would take care of this peculiar initial condition.

One finds that a period of inflation(3,4), with constant energy density with its consequent exponential increase of the scale parameter, $R(t) \sim e^{H_0 t}$ "solves" this type of problem provided that $H_0 \delta t \geq 60$. However such a scenario makes the K/R^2 term in equation (1) totally negligible. Thus inflation predicts that today Ω should be equal to 1, and yet it is measured to be at most 0.3! The nature of the missing matter is not clear at present (if it exists at all), but there is no lack of candidates from particle physics model.

At this stage, we should mention one great advantage of the inflationary scenario: it provides the model builder with a natural mechanism to get rid of embarrassing particles. All that one has to do is to arrange for the unwanted particle not to be produced after inflation takes place, via the so-called reheating process. This trick, as we shall see, is used at least twice, once for monopoles, once for gravitinos. This type of procedure was known to the ancients (re: Gilgamesh and/or Noah's ark).

Let us now describe sensible requirements on the SQID. First of all the scale of GUT breaking, M_x, is bounded by the absence of proton decay to be greater than 10^{15} GeV or 1 milliPlanck. Such a breaking produces magnetic monopoles, which do not appear to be plentiful in nature. Since it seems that monopoles cannot be destroyed except by antimonopoles, one had better arrange for inflation to take place after GUT breaking. In this way, the monopole density will be diluted by the exponential phase (inflation). However in the reheating after inflation the germs of baryogenesis must appear, and this poses very severe constraints on the various models.

Finally it is desirable to include supersymmetry (Susy "pour les intimes") in order to protect the weak interaction scale[5]. This implies a typical splitting within a supersymmetric multiplet $\mathcal{O}$(Tev). If μ is the SUSY breaking scale, then the gravitino mass $m_{\tilde{G}}$ is given by

$$m_{\tilde{G}} \sim \frac{\mu^2}{M} \sim \text{Tev} \qquad (5)$$

i.e. $\mu \sim 10^{10}$ GeV, or 10 nanoPlanck. The gravitino is a particle which is itself embarrassing. For this particular mass, it would decay after cosmological nucleosynthesis. Its decay would have several negative consequences[6] - one it will produce more photons, thus reheating and then diluting the baryon asymmetry, second its decay products would change the abundance of elements laboriously formed earlier. For instance in the decay $G \to \gamma + \tilde{\gamma}$ ($\tilde{\gamma}$ is the photino), photons capable of dissociating deuterium would be produced. On the other hand in the decay into a gluon-gluino pair, $\bar{p}$ would eventually result, which could change ^4He into D. The only way out seems to be to limit the original abundance of gravitinos. Thus can be done by appealing to inflation. Then one has to worry only about gravitinos produced either by reheating or by the decay of particles whose interactions break supersymmetry. The first constraint puts a limit to the temperature of the reheated bath at $T_R \lesssim 10^8 - 10^{12}$ GeV, depending on who you read. The second constraint (called the Entropy Crisis of Supersymmetry) has to be examined for each model of SUSY breaking.

Thus we will need three mechanisms for our various constraints: in temporal order 1) GUT breaking at a scale M_x ($\sim$mP), Inflation at a scale Δ, and SUSY breaking at a scale μ ($\sim 10^{x}$ nP). We now present our model[7]. We will use the formalism of N=1 supergravity to describe it.

Starting from a chiral superfield $\Phi_i(x,\theta)$, each containing a complex spinless boson and a Weyl fermion, one describes their non gauge interactions by means of a superpotential $P(\Phi)$, from which the ordinary potential is derived[8]

$$V(\phi_i) = e^{-\frac{|\phi_i|^2}{M^2}} \{\sum_i | \frac{\partial P}{\partial \Phi_i} + \frac{\phi_i^* P}{M^2} |^2 - \frac{3}{M^2} |P|^2\}, \qquad (6)$$

where ϕ_i are the spinless components of Φ_i. This form is valid when the scalar fields have the usual kinetic terms. This potential breaks SUSY if and only if

$$|\frac{\partial P}{\partial \Phi} + \frac{\phi^* P}{M^2}| \neq 0 \text{ at minimum.} \qquad (7)$$

Now if one starts from a superpotential made up of the sum of several disconnected parts, the structure of (6) forces interactions between them of $\mathcal{O}(\frac{1}{M^2})$. The philosophy of hidden sector physics is to have different sectors which interact with one another only through gravitational strength interactions.

Our model for the superpotential is then

$$P = I + G + S, \qquad (8)$$

where I describes inflation, S describes supersymmetry breaking, and G describes the GUT sector. Thus in this picture fields in a given sector will not be in thermal equilibrium with fields of another sector until very late times when the expansion rate is very slow. This is because they only interact gravitationally with one another. [Interestingly, in a contracting universe, particles would be thrown together and equilibrium would be easier with time, but this does not apply to our universe, at least not yet.]

The I part of the superpotential is now built according to the following tenets - use only one superfield containing one complex scalar field, the inflaton, and its partner the inflatino - at minimum its potential does not generate any cosmological term and does not break supersymmetry, i.e.

$$\left|\frac{\partial I}{\partial \phi} + \frac{\phi^* I}{M^2}\right|^2 - \frac{3}{M^2}|I|^2 = 0 \qquad \text{at minimum,} \tag{9}$$

and

$$\left|\frac{\partial I}{\partial \phi} + \frac{\phi^* I}{M^2}\right| = 0 \qquad \text{at minimum.} \tag{10}$$

These together imply that both I and $\frac{\partial I}{\partial \phi}$ are zero at minimum. Hence the simplest expression for I obeying these criteria is

$$I = \frac{\Delta^2}{M}(\Phi - \phi_0)^2 , \tag{11}$$

where Δ is an unknown parameter with dimension of mass and ϕ_0 is the vacuum value of the inflaton. We require further that I inflates, i.e. $\frac{\partial V}{\partial \phi} = 0$ at the origin. In this case we find that ϕ_0 is fixed to be M. Hence we set

$$I = \frac{\Delta^2}{M}(\Phi - M)^2. \tag{12}$$

As an added bonus we find that $\frac{\partial^2 V}{\partial \phi^2} = 0$ at the origin as well. In fact near the origin

$$V_I(\phi) \approx \Delta^4[1 - (\phi/M)^3 + ..]; \quad \phi/M \ll 1, \tag{13}$$

where ϕ is the inflaton field. The inflaton potential has a minimum at ϕ=M, where the mass of the inflaton field is

$$m_\phi = \frac{\Delta^2}{M} . \tag{14}$$

This very simple form for the inflaton sector leads to the following scenario. The inflaton has only gravitational strength self interactions. One can check that except for temperatures near the Planck mass it never is in thermal equilibrium. It produces about 10^8 e-folds of inflation. At $t_* \sim m_\phi^{-1}$ the inflaton field remembers its mass and inflation stops; the evolution becomes matter dominated until a time $t_R \sim \Gamma_\phi^{-1}$ when the inflaton decays. It decays universally since it is hidden from all the other sectors. Its decay rate is given by

$$\Gamma_\phi \sim \frac{\Delta^6}{M^5} , \tag{15}$$

corresponding to a reheat temperature

$$T_R \sim (M\Gamma_\phi)^{1/2} \sim \frac{\Delta^3}{M^2} . \tag{16}$$

Now the hitherto unknown parameter Δ is fixed by the needed scale of density fluctuations; these are given by[9]

$$\frac{\delta\rho}{\rho} \simeq \frac{H^2}{\dot{\phi}} , \tag{17}$$

evaluated at a time during inflation when the fluctuation leaves the horizon. In our model one can estimate this expression for fluctuation relevant to galactic sizes to be

$$\frac{\delta\rho}{\rho} \simeq 1.5 \times 10^4 \left(\frac{\Delta}{M}\right)^2 , \tag{18}$$

by using the approximation (13) for the potential, which is certainly valid in this regime. Thus in order to get the desired value of $\frac{\delta\rho}{\rho} \simeq 10^{-4}$, we must fix

$$\Delta \simeq 10^{-4}\ P \approx .1 \text{ milliPlanck}. \tag{19}$$

There are no more free parameters left in this sector. Let us now turn to the other sectors and see if contradictions can be found.

We take the SUSY breaking superpotential to be of the O'Raifeartaigh[10] type, i.e.

$$S = A[\lambda B^2 + \mu^2] + \mu BC + q, \tag{20}$$

where A is the superfield whose F-term breaks the supersymmetry. Its scalar component is called, the "O'Raifearton" and its spinor component is "eaten" by the gravitino to provide the helicity ± 1/2 states. The O'Raifearton acquires a mass at the one loop level

$$m_A \simeq \alpha_\lambda \mu \ ; \quad \alpha_\lambda = \frac{\lambda^2}{4\pi} , \tag{21}$$

and its preferred decay channel is into gravitinos

$$A \to \tilde{G}\tilde{G} \ , \quad \Gamma_A \sim \frac{\alpha_\lambda^3}{(4\pi)^2} m_A \tag{22}$$

Note that with our value of Δ, the inflaton is (barely) kinematically forbidden to decay into the O'Raifearton, since

$$m_\phi \lesssim m_A \tag{23}$$

Alternatively, we could say that (23) puts an upper bound on Δ, specifically $\Delta \lesssim 10^{-3.5}$ P, perfectly consistent with the value (19) obtained from fluctuations.

Next one checks that as long as α_λ is reasonably small the decay of the inflaton into O'Raifearton by one loop effects can also be suppressed. Thus in our model, there is essentially no direct production of gravitinos by inflaton decay. However, there is a more worrisome source of O'Raifeartons which comes about by means of the induced interactions between I and S: after inflation, the A field is left with a non-zero potential energy. This potential energy is then eventually translated into gravitinos. However one can check that with our values for μ and Δ this produces only a negligible gravitino abundance at the time of nucleosynthesis. Further since our reheat temperature is, for our value of Δ, low i.e. $\sim 10^6$ GeV, we completely avoid the gravitino problem.

There remains, however, one hurdle to overcome, and it is that of baryogenesis[2]. One wants the reheating process to include particles whose interactions can lead to baryon asymmetry; yet the inflaton mass is fixed by Δ [$M_\phi \sim 10^{10}$ GeV] - it follows that the Higgs triplets generated in inflaton decay cannot have masses higher than that, otherwise baryogenesis could not occur. On the other hand such low mass for the triplets will cause in principle fast proton decay. Thus we are in the enviable (but dangerous) situation of obtaining an <u>upper</u> bound for proton decay[11] from the above cosmological considerations!

Because the inflaton mass is bounded in our model, we have to consider Higgs triplets with masses $\mathcal{O}(10^{10}$ GeV). Also because of the low reheat temperature (10^6 GeV), we have to rely on a non-equilibrium scenario for producing baryon asymmetry. Fortunately the inflaton is never in equilibrium, so we can check that its preferential decay modes will occur into Higgs triplets. We have then[12]

$$n_{B-\overline{B}} \simeq n_H \delta B \approx n_\phi \delta B, \tag{24}$$

where δB is the specific baryon number asymmetry, n_H is the triplet density, n_ϕ the inflaton density. Also then

$$n_\gamma T_R \approx m_\phi n_\phi, \tag{25}$$

so that

$$\frac{n_{B-\overline{B}}}{n_\gamma} \simeq \frac{\Delta}{M} \delta B \; . \tag{26}$$

There remains to see if we can arrive at GUTs models which produce the required δB while not causing too fast a proton decay rate.

In a supersymmetric GUT, the presence of Higgs triplets with mass $\mathcal{O}(10^{10}$ GeV)(13) can cause fast proton decay by means of the quark chiral operators (dimension five). Specifically given the Yukawa couplings

$$\phi_{\bar{5}}\phi_{10}H_1 + \phi_{10}\phi_{10}H_2 \, , \tag{27}$$

where H_1 and H_2 contain the Higgs color triplets, the following interactions are induced

a) $\phi_{\bar{5}}\phi_{10}\bar{\phi}_{\bar{5}}\bar{\phi}_{10}$, mediated by H_1

b) $\phi_{10}\phi_{10}\bar{\phi}_{10}\bar{\phi}_{10}$, mediated by H_2

and

c) $\phi_{\bar{5}}\phi_{10}\phi_{10}\phi_{10}$, mediated by H_1 and H_2

exchange, provided H_1 and H_2 mix. The last interaction contributes to proton decay in supersymmetric theories as an interference term, and for $H_1, H_2 \sim \mathcal{O}(10^{10}$ GeV) with $\mathcal{O}(1)$ mixing, will cause fast proton decay, in contradiction with experiment. The only hope is to suppress the mixing between H_1 and H_2, while relying on straight H_1 and H_2 exchange to give proton decay. A careful analysis of these decay rates already exists in the literature(14). One finds that if H_1 is lighter than H_2, the preferred decay modes are $p \to e^+\pi^0$, $e^+\omega$ and our limit from the inflaton mass gives(11)

$$\tau_p(p \to e^+\pi^0(\omega)) \lesssim 10^{34} \text{ years;} \tag{28}$$

it is therefore uninteresting since known backgrounds make this upper limit unattainable. However, if H_2 is lighter than H_1 the limit becomes more interesting, namely(11)

$$\tau_p(p \to \mu^+K^0) \lesssim 10^{31} \text{ years.} \tag{29}$$

The next question involves the actual mechanism by which the baryon asymmetry is produced. In our model we have assumed that the $H_1 - H_2$ mixing term is suppressed. Such models exist(11) but they seem to involve a large number of Higgs particles.

At any rate we hope that we have shown that with a definite cosmological model for inflation, the existence of baryogenesis and the absence of proton decay poses severe constraints on any SQID one might want to propose.

Acknowledgement

The author thanks the Lewes Center for Physics for providing a suitable environment for writing this paper.

References

(1) See one of the many excellent books on the subject; e.g. S. Weinberg, Gravitation and Cosmology, New York, Wiley 1972.

(2) M. Yoshimura, Phys. Rev. Lett. 41, 281 (1978), (E) 42, 746 (1978). A. D. Sakharov, Zh. Eksp. Teor. Fiz. Pis'ma 5, 32 (1967). E. Kolb, M. Turner in Ann. Rev. Nucl. Part. Sci. 1983, 33:645.

(3) A. Guth, Phys. Rev. D23, 347 (1981).

(4) A. D. Linde, Phys. Lett. 108B, 389 (1982); A. Albrecht and P. J. Steinhardt, Phys. Rev. Lett. 48, 1220 (1982). P. J. Steinhardt and M. S. Turner, Phys. Rev. D29 2126 (1984).

(5) L. E. Ibañez, Phys. Lett. 118B, 73 (1982); R. Barbieri, S. Ferrara and C. A. Savoy, Phys. Lett. 119B (1982) 343; P. Nath, R. Arnowitt and A. H. Chamseddine, Phys. Rev. Lett. 49 (1982) 970; H. P. Nilles, M. Srednicki and D. Wyler, CERN preprint TH-3432 (1982); S. Ferrara, D. V. Nanopoulos and C. A. Savoy, Phys. Lett. 123B (1983) 214; L. Hall, J. Lykken and S. Weinberg, University of Texas preprint UTTG-1-83 (1983); L. E. Ibañez and G. G. Ross, Phys. Lett. 131B (1983) 335.

(6) For the latest see M. Yu. Khlopov and A. D. Linde, Phys. Lett. 138B, 265 (1984); John Ellis, Jihn E. Kim and D. V. Nanopoulos, CERN preprint 3839 (1984).

(7) R. Holman, P. Ramond and G. G. Ross, Phys. Lett. 137B, 343 (1984); G. D. Coughlan, R. Holman, P. Ramond, G. G. Ross, University of Florida preprint, 1984; R. Holman, University of Florida preprint UFTP-84-6 (1984).

(8) E. Cremmer et al., Phys. Lett. 79B, 231 (1978); Nucl. Phys. B147, 105 (1979); Phys. Lett. 116B, 231 (1982); Nucl. Phys. B212, 413 (1982).

(9) See for instance J. M. Bardeen, P. J. Steinhardt and M. S. Turner, Phys. Rev. D28, 679 (1983) and references contained within.

(10) L. O'Raifeartaigh, Nucl. Phys. B96, 331 (1975).

(11) G. Coughlan, R. Holman, P. Ramond and G. G. Ross, in preparation.

(12) L. F. Abbott, E. Fahri, and M. B. Wise, Phys. Lett. 117B, 29 (1982).

(13) B. Grinstein, Nucl. Phys. B206, 387 (1982); S. Dimopoulos and F. Wilczek, ICTP UM-HE81-71 (1982); A. Masiero et al., Phys. Lett. 117B, 380 (1982).

(14) P. Salati and J. C. Wallet, Nucl. Phys. B209, 389 (1982).

GENERATION AND EVOLUTION OF ENERGY DENSITY FLUCTUATIONS IN INFLATIONARY UNIVERSE MODELS

Robert H. Brandenberger

Institute for Theoretical Physics
University of California
Santa Barbara, California 93106

ABSTRACT

Spatial correlations in the vacuum state wave functional of a scalar quantum field generate classical matter perturbations in the de Sitter phase of inflationary universe models. The primordial energy density fluctuation spectrum at initial Hubble radius crossing is scale invariant. We analyze the evolution of perturbations outside the Hubble radius, demonstrate that the spectrum remains independent of scale when measured at the time the scales reenter the present horizon, and give a general formula to determine the amplitude of the fluctuation spectrum for a wide class of particle physics models.

1. Introduction

Inflationary universe models[1),2)] provide a causal mechanism which from first principles generates the primordial energy density fluctuations required as initial conditions in theories of galaxy formation. We will discuss how quantum fluctuations in the matter fields give rise to classical matter perturbations, and how these classical inhomogeneities evolve gravitationally outside the Hubble radius.

We begin by giving a brief review of the cosmology of inflationary universe models.[1),2)] The main features can be discussed using Fig. 1. Inflationary universe models are characterized by an initial de Sitter phase, a phase in which the constant vacuum energy of the quantum field acts as an effective cosmological constant and leads to an exponential growth of the scale factor, $a(t) \sim \exp(tH)$. At reheating time t_R the vacuum energy is transformed into thermal energy of particles. Hence for $t > t_R$ the universe becomes a radiation dominated Friedmann-Robertson-Walker (FRW) universe and $a(t) \sim t^{\frac{1}{2}}$. (We are restricting our attention to spatially flat Robertson-Walker metrics of the form

$$ds^2 = - dt^2 + a(t)^2 \, d\underline{x}^2 \text{ .)} \tag{1}$$

The first curve on Fig. 1 shows the physical distance of the Hubble radius $H^{-1}(t)$ as a function of time $(H(t) = \dot{a}(t)a(t)^{-1})$. The Hubble radius is the maximal distance over which microphysics can act coherently. A simple heuristic way to understand this is to note that for plane wave solutions of the equation of motion of a scalar field, spatial gradient terms are negligible if the wavelength exceeds the Hubble radius. We are interested in the location of matter perturbations relative to the Hubble radius. Matter perturbations have constant comoving coordinates. Hence their physical distance from a comoving observer evolves like the scale factor $a(t)$.

Perturbations on scales of cosmological interest, i.e. on the scales of galaxies or clusters of galaxies have only recently entered the Hubble radius in the FRW era. In inflationary universe models, however, as plotted in Fig. 1, they originate inside the Hubble radius in the de Sitter phase. Therefore it is possible to hope for a causal generation mechanism. This is the first major success of the new cosmological scenarios with regards to a theory of the origin of fluctuations.

The shape of the theoretically predicted spectrum constitutes the second major success: the amplitude of fluctuations is independent of the comoving wave number k when measured at the time $t_f(k)$ when the scale enters the Hubble radius. This scale invariant Zeldovich spectrum was postulated many years ago to explain galaxy formation in the context of the adiabatic fragmentation picture.[3)] The heuristic reason for this result is simple.[4)] Consider perturbations on two different scales k_1 and k_2. Since the de Sitter phase is time translation invariant and since the evolution of the two perturbations up to the time $t_i(k)$ when the scales leave the Hubble radius are related by time translation, the amplitude of fluctuations at initial Hubble radius crossing $t_i(k)$ will be scale invariant. Outside the Hubble radius microphysics cannot act coherently. Thus the "physical" size of perturbations must remain unchanged, and a scale invariant spectrum will result also at (final) Hubble radius crossing $t_f(k)$.

The above heuristic arguments for a scale invariant Zeldovich spectrum are very suggestive, but by no means conclusive. If we imagine imposing initial conditions on a constant time hypersurface early in the de Sitter phase, we immediately break the time translational symmetry of the problem. In the first part of this lecture we prove that quantum fluctuations in the de Sitter phase of a cosmological model indeed generate classical matter perturbations with a spectrum which is scale invariant at initial Hubble radius crossing.

Most measures of the magnitude of classical gravitational fluctuations do not remain constant while the scales are outside the Hubble radius. From Fig. 1 it is obvious that the time interval during which a perturbation is outside the Hubble radius depends on the scale. Hence it does not seem obvious at first that the spectrum at final Hubble radius crossing $t_f(k)$ will be scale invariant. In the second part of this lecture we show that the factor by which perturbations are amplified between $t_i(k)$ and $t_f(k)$ depends only on the change in the equation of state. In particular it is independent of k, and does not depend on the details of the reheating mechanism (this section summarizes joint work with Ronald Kahn[5]).

For any comoving wave number k there are two completely different time regions. For $t < t_i(k)$, perturbations on scale k are inside the Hubble radius, microphysics can act coherently, and quantum fluctuations in the matter fields will generate classical energy density fluctuations. For simplicity we consider a single scalar field. Two approximations will be important during the period $t < t_i(k)$. First, we neglect metric perturbations, motivated by our objective to study the origin of matter perturbations in an initially homogeneous universe. The approximation is self-consistent since the gravitational inhomogeneities induced by the matter perturbations via the Einstein constraint equations remain very small up to $t_i(k)$. The second approximation is to consider a free scalar field instead of a scalar field with quantum field nonlinearities given by the inflationary potential of Fig. 2. Since in new inflationary universe models[2] the potential near the origin is virtually flat and since the scalar field $\phi_0(t)$ remains in the flat section of the potential until long after all scales of cosmological interest have crossed the Hubble radius, the approximation should be reasonable.

For $t_i(k) < t < t_f(k)$ the perturbation is outside the Hubble radius and microphysics can no longer act coherently. Therefore, the fluctuations set up before $t_i(k)$ will only feel gravitational forces and will evolve according to the classical Einstein equations linearized about the background solution.

2. Generation of Perturbations

In inflationary universe models the matter source for general relativity is taken to be a quantum field theory, more precisely a grand unified field theory with a (gauge) symmetric ground state with a large vacuum energy of the order σ^4 (σ is the scale of grand unification). The physics of the generation and evolution of perturbations can be discussed in a toy model of a single scalar field with potential sketched in Fig. 2. As mentioned above the free field theory approximation is well justified in the period $t < t_i(k)$ in which the perturbations are formed.

The basic question is how inhomogeneities can develop given a homogeneous and isotropic background geometry and a quantum state which does not break space translational invariance. The mechanism must be nontrivial as the following incorrect analysis shows: The Einstein equations are

$$G_{\mu\nu} = 8\pi G T_{\mu\nu}^{cl} \quad . \tag{2}$$

What is the classical energy-momentum tensor? The naive answer would be to take the expectation value of the quantum operator $T_{\mu\nu}$ in the given quantum state $|s\rangle$, i.e.

$$T_{\mu\nu}^{cl}(\underline{x},t) \equiv \langle s|T_{\mu\nu}(\underline{x},t)|s\rangle . \tag{3}$$

But since the state is space-translation invariant, the expectation value of $T_{\mu\nu}$ will be space-independent. The naive approach thus predicts the absence of fluctuations.

It is not hard to see what is missing in the above argument: although the state $|s\rangle$ is space-translationally invariant, it contains spatial correlations. The main point of this section is to demonstrate how spatial correlations give rise to classical matter fluctuations.

The physical argument which predicts nonvanishing perturbations is based on Hawking radiation.[6] An observer with a particle detector moving along a world line in de Sitter space[7] or in the de Sitter phase of a FRW universe[8] will detect a thermal flux of particles corresponding to the temperature

$$T_H = \frac{H}{2\pi} . \tag{4}$$

It is, however, incorrect to conclude that there will be thermal matter fluctuations. After all, the energy-momentum tensor of Hawking radiation is not that of a thermal bath ($p \neq \frac{1}{3}\rho$). It is de Sitter invariant[9] ($p = -\rho$) or at least (in the case of an approximate de Sitter phase of a FRW universe) approximately de Sitter invariant.[10]

The shape of the primordial fluctuation spectrum is usually obtained by applying the correlation function method. The mean deviation of a stochastic function $f(\underline{x})$ on wave number $\underline{k}$ is related to its autocorrelation function by

$$(\delta f(k))^2 = \frac{k^3}{(2\pi)^3} \int d^3\underline{x}\, e^{i\underline{k}\cdot\underline{x}} \langle f(\underline{x})f(\underline{0})\rangle . \tag{5}$$

The authors of Ref. 11 obtain initial fluctuations in the classical matter field $\phi(\underline{x},t)$ (which couples to gravity) by applying Eq. (5) and using the two-point Green's function in place of the classical autocorrelation function.

We will now justify the correlation function method from "first principles." Quotation marks are necessary since in any semiclassical approach to coupling matter to gravity, i.e. an approach in which matter is quantized but gravity is treated classically, the prescription for the classical energy-momentum tensor of Eq. (2) must be put in by hand.

To motivate the physical arguments for a scalar quantum field theory, we first consider a quantum mechanical harmonic oscillator with one degree of freedom q in an expanding universe. The ground state wave function is a Gaussian with width inversely proportional to the frequency of the oscillator. In an expanding universe the frequency scales as $a(t)^{-1}$. Hence the width of the wave function increases as a(t). The r.m.s. distance of the wave packet gives the classical value q^{cl} of the harmonic oscillator. The r.m.s. distance is given by the two-point function. Hence in a state $|\psi\rangle$

$$q^{cl}(t) = (\langle\psi|Q^2|\psi\rangle)^{1/2} . \tag{6}$$

A scalar quantum field theory is a quantum system with an infinite number of degrees of freedom. A state wave functional contains not only information about one degree of freedom, but also about correlations between different degrees of freedom, i.e. spatial correlations. These correlations correspond to classical field perturbations. Thus the basic idea of our approach[12] is to define a classical field $\phi_{cl}(\underline{x},t)$ which contains information about both the r.m.s. value of any single degree of freedom and the spatial correlations in the vacuum state, the vacuum fluctuations.[13] We define

$$\phi_{cl}(\underline{x},t) \equiv \phi_0(t) + \delta\phi(\underline{x},t) . \tag{7}$$

$\phi_0(t)$ gives the spread of the vacuum state wave functional. Hence in analogy to the case of Eq. (6) for a single degree of freedom

$$\phi_0(t) \equiv (\langle\psi_0|\Phi^2(\underline{x})|\psi_0\rangle)^{1/2} . \tag{8}$$

We will work in the Schrödinger representation in which the state wave functionals carry the time dependence. $|\psi_0\rangle$ denotes the vacuum state. Spatial correlations manifest themselves in nonvanishing expectation values of the operator $\tilde{\Phi}(\underline{k})^2$. Hence we define $\delta\phi$ in momentum space by

$$\begin{aligned}\delta\tilde{\phi}(\underline{k},t)^2 &\equiv \langle\psi_0||\tilde{\Phi}(\underline{k})|^2|\psi_0\rangle \\ &= V\int d^3\underline{r}\; e^{i\underline{k}\cdot\underline{r}} \langle\psi_0|\Phi(\underline{r})\Phi(\underline{0})|\psi_0\rangle .\end{aligned} \tag{9}$$

V is a volume cutoff. In terms of the two-point function in momentum space

$$\langle\psi_0|\tilde{\Phi}(\underline{k})^*\tilde{\Phi}(\underline{l})|\psi_0\rangle = \frac{(2\pi)^3}{V}\,\delta^3(\underline{k}-\underline{l})\,\delta\tilde{\phi}^2(\underline{k},t) . \tag{10}$$

We can now use the standard formulas for the energy-momentum tensor of a classical scalar field to compute $T_{\mu\nu}{}^{cl}(\underline{x},t)$ in terms of the classical field $\phi_{cl}(\underline{x},t)$. In particular, the classical energy density fluctuation becomes

$$\frac{\delta\tilde{\rho}}{\rho}(\underline{k},t) = \phi_0(t)\,\delta\phi(\underline{k},t)\,\rho^{-1} . \tag{11}$$

ρ is the vacuum energy of the theory.

Our ideas lead to the following procedure for computing initial energy density fluctuations at Hubble radius crossing $t_i(\underline{k})$. We consider a free massless scalar field in the de Sitter phase of the cosmological model. The first step is to define what we mean by the vacuum state (we will return to this point). We then compute the two-point functions in position and momentum space, use Eqs. (7)-(10) to define a classical field which is not homogeneous in space, and finally evaluate Eq. (11) at $t_i(\underline{k})$ to obtain the energy density fluctuations at initial Hubble radius crossing.

The calculation is described in a separate publication.[12)] Here we will only elaborate on some of the important technical points, the most important of which is the correct definition of the vacuum state. The vacuum $|\psi_0\rangle$ is defined by

$$a_{\underline{k}}|\psi_0\rangle = 0 \qquad \forall\, \underline{k} \tag{12}$$

where the $a_{\underline{k}}$ are the annihilation operators, operator coefficients of the expansion of the quantum field $\Phi(\underline{x},t)$ in terms of positive and negative frequency solutions $g_{\pm}^{\underline{k}}(t)\exp(i\underline{k}\cdot\underline{x})$ of the classical field equations:

$$\Phi(\underline{x},t) = (2\pi)^3 \int d^3\underline{k}\{a_{\underline{k}} g_+^{\underline{k}}(t)\; e^{i\underline{k}\cdot\underline{x}} + \text{h.c.}\} \tag{13}$$

(h.c. stands for Hermitean conjugate). The prescription relies on having a predetermined space-time slicing, a time direction in order to define positive and negative frequencies and a flat spatial section in order to be able to Fourier transform. In special relativity there is a unique class of inertial frames; in general relativity there is not. Hence for quantum field theories in curved space-time there is no unique vacuum state. On the other hand, given any frame we can define a state by the above second quantization prescription. The state obtained is empty of scalar particles when viewed from the given frame.

Our choice of frame is motivated by physical considerations. In the usual inflationary universe models a hot FRW period precedes the de Sitter phase. Matter is a thermal bath of relativistic particles in the FRW phase. The bath is at rest with respect to the FRW coordinate frame. After a few Hubble times into the de Sitter phase the particles initially present will have redshifted away exponentially and the resulting state can be described as empty of particles in the coordinate frame. Hence our vacuum state is defined by Eq. (12) in the FRW coordinate frame.

We calculate the vacuum state wave functional $\psi_0(\phi(\underline{x}),t)$, the quantum field theory analog of the ground state wave function in quantum mechanics, defined by the formal expansion

$$|\psi_0\rangle = \int [d\phi]\psi_0(\phi(\underline{x}),t)|\phi(\underline{x})\rangle \;. \tag{14}$$

$|\phi(\underline{x})>$ is an eigenstate of the field operator $\Phi(\underline{x})$ with eigenvalue $\phi(\underline{x})$; the functional integral runs over all classical field configurations.

Our result is not surprising. The ground state wave function of a quantum mechanical harmonic oscillator is a Gaussian with width inversely proportional to its frequency. A free scalar field theory is an infinite assembly of uncoupled harmonic oscillators, one for each wave number $\underline{k}$. Hence we expect the ground state wave functional to be the product of the harmonic oscillator wave functions for each $\underline{k}$. The effect of curved space-time is to render the frequencies time dependent. The result is[12)]

$$\psi_0(\phi(\underline{k}),t) = N \exp\left\{-\frac{1}{2} a^3(t)(2\pi)^{-3}\int d^3\underline{k}\omega(\underline{k},t)|\phi(\underline{k})|^2\right\} . \tag{15}$$

N is the normalization constant and $\omega(\underline{k},t)$ is the frequency of the oscillator labeled by comoving coordinate $\underline{k}$. $\omega(\underline{k},t)$ will depend on the solutions $g_+^{\underline{k}}(t)$ which in turn depend on the coupling of gravity to matter. For minimal coupling (coefficient of the $R\Phi^2$ term in the Lagrangian of the quantum field set to zero, R being the Ricci scalar) the solutions are

$$g_+^{\underline{k}}(t) = [1 - i(-k\tau)]\, e^{-ik\tau} \tag{16}$$

where $\tau(t) = - H^{-1}(-Ht)$ is conformal time. Hence

$$\mathrm{Re}\ \omega(\underline{k}) = ke^{-tH}\frac{(-k\tau)^2}{1+(-k\tau)^2} . \tag{17}$$

Since the vacuum state wave functional is a Gaussian in momentum space, the two-point function in momentum space is simply

$$<\psi_0|\tilde{\Phi}(\underline{k})^*\Phi(\underline{1})|\psi_0> = \delta^3(\underline{k}-\underline{1})(2\pi)^3 a^{-3}(t)\,\frac{1}{2}\,[\mathrm{Re}\ \omega(\underline{k})]^{-1} . \tag{18}$$

The two-point function in position space can be expressed as a Fourier transform of the momentum space Green's function:

$$<\psi_0|\Phi(\underline{x})\Phi(\underline{x}+\underline{r})|\psi_0> = (2\pi)^{-2}a^{-3}(t)\,\frac{1}{r}\int_0^\infty dk\ k\ \sin kr[\mathrm{Re}\ \omega(\underline{k})]^{-1} . \tag{19}$$

The integral is both infrared and ultraviolet divergent. The infrared divergence is eliminated by imposing a cutoff $k_{min} = H$. This means that perturbations on wavelengths larger than the Hubble radius at the beginning of the de Sitter phase are set to zero. Since such perturbations are given by the initial conditions at the big bang rather than by quantum processes in the de Sitter phase, it makes sense to consider them as contributing to a local background. The ultraviolet divergence is handled by renormalization.[12),13)] For minimal coupling we recover the result of Vilenkin and Ford[14] and Linde[15]:

$$\langle\psi_0|\Phi^2(\underline{x})|\psi_0\rangle = (2\pi)^{-2}H^3 t \ . \tag{20}$$

For non-minimal coupling there is an effective gravitationally-induced potential barrier (stemming from the $R\Phi^2$ term in the Lagrangian) which keeps the expectation value of $\Phi^2(\underline{x})$ constant in time. For minimal coupling, however, the wave functional will spread.

The physical interpretation of the two-point function in momentum space is interesting. The Green's function scales as in Minkowski space for wavelengths inside the Hubble radius ($-k\tau > 1$) but like in de Sitter space for wavelengths with ($-k\tau < 1$).

The Green's functions (18) and (20) determine the classical field. By Eqs. (8) and (9)

$$\phi_0(t) = (2\pi)^{-1}H^{3/2}t^{1/2} \tag{21}$$

and

$$\delta\phi(\underline{k},t) = V^{1/2}(2\pi)^{3/2}a^{-3/2}(t)\,\frac{1}{\sqrt{2}}\,[\mathrm{Re}\ \omega(\underline{k})]^{-1/2} \ . \tag{22}$$

Hence by Eq. (11), at initial Hubble radius crossing $t_i(\underline{k})$

$$\frac{\delta\rho}{\rho}\,(\underline{k},t_i(\underline{k})) = O(1)V^{1/2}k^{-3/2}\left(\frac{H}{\sigma}\right)^4 . \tag{23}$$

The crucial factor $k^{-3/2}$ stems from evaluating $a(t)$ at Hubble radius crossing $H^{-1} = k^{-1}a(t_i(\underline{k}))$. ρ is of order σ^4 and since H is the only dimensional parameter in the quantum problem, the factor H^4 is required by dimensional arguments (it is easy to explicitly trace its origin).

Equation (23) is our final result. It represents a scale invariant Zeldovich spectrum with amplitude $(H/\sigma)^4$ at initial Hubble radius crossing. We recall that the quantity which characterizes the physical size of perturbations on a length scale k^{-1} is the average mass excess $\delta M/M$ inside a ball of radius k^{-1}. Fluctuations in the energy density on scales smaller than k^{-1} average to zero in the ball and hence do not contribute to $\delta M/M$ on scale k^{-1}. Thus[16]

$$\left(\frac{\delta M}{M}\right)^2(k) = O(1)V^{-1}k^3\left(\frac{\delta\rho}{\rho}\right)^2(k) \ , \tag{24}$$

3. Classical Propagation of Fluctuations Outside the Hubble Radius

Once outside the Hubble radius microphysics can no longer influence perturbations, and the latter will hence evolve according to gravity alone. Since the inhomogeneities at $t_i(k)$ are very small (in typical grand unified theories H/σ is of the order 10^{-5}), the growth of perturbations can be described by linear perturbation theory. Most convenient is the gauge invariant formalism established by Bardeen.[17),18)] We consider only scalar metric fluctuations since these are the only ones which couple to energy density and pressure perturbations.

The complete dynamical system consists of three parts: the homogeneous and isotropic background metric described by the scale factor $a(t)$, the matter -- in our case the scalar field $\phi(\underline{x},t)$ -- and finally the scalar metric perturbations. The most general scalar metric perturbation of the background FRW metric of Eq. (1) can be expressed in terms of four free functions $A(\underline{x},t)$, $B(\underline{x},t)$, $E(\underline{x},t)$, and $F(\underline{x},t)$

$$\delta g_{\mu\nu} = a^2(t) \begin{pmatrix} E & F_{,i} \\ F_{,i} & A\delta_{ij} + B_{,ij} \end{pmatrix} \qquad (25)$$

This description is redundant. We are only interested in the physical degrees of freedom, not in those corresponding to gauge transformations, transformations of the background coordinates. As shown in Refs. 17 and 18, a single gauge invariant function contains all physical information about scalar metric perturbations. This function Φ_H is given by

$$2\Phi_H = A - a\dot{a}\dot{B} + 2a\dot{a}F \ . \qquad (26)$$

Φ_H is mathematically gauge invariant; it takes on a concrete physical meaning in a specific gauge. For our purposes the only important fact is that at horizon crossing t_H, Φ_H equals the energy density fluctuation in comoving gauge (denoted by subscript c)

$$\Phi_H(t_H) = \frac{3}{2}\left(\frac{\delta\rho}{\rho}\right)_c (t_H) \ . \qquad (27)$$

We use Eq. (27) to determine the metric perturbations at initial Hubble radius crossing $t_i(\underline{k})$ generated by the energy density fluctuations discussed in the previous section. We then follow the dynamical evolution of Φ_H until the scales reenter the Hubble radius at $t_f(\underline{k})$ and, by Eq. (27), obtain $\delta\rho/\rho$ at that time. Our analysis (joint work with Ronald Kahn[5)]) confirms the findings of previous investigations[11)] that the spectrum at $t_f(\underline{k})$ is scale invariant. We show that the final amplitude depends only on the change in the equation of state between initial and final Hubble radius crossing and, in particular, is insensitive to the details of the reheating mechanism. In the following we will present the main physical ideas of the analysis.

The equation of motion for Φ_H is obtained by linearizing the Einstein equations about the FRW background. The idea is simple, but not the technical details.[17),18)] The result shows that the evolution of Φ_H depends only on the equation of state of the background:

$$\ddot{\Phi}_H + (4 + 3c_s^2)H\dot{\Phi}_H + 3(c_s^2 - w)H^2\Phi_H = I(t) . \tag{28}$$

$w = p/\rho$ is the ratio of pressure to energy of the background. w and $c_s^2 = \dot{p}/\dot{\rho}$ (c_s is not, in general, the speed of sound) describe the equation of state of the background; $I(t)$ is a gauge invariant combination of matter source terms and is negligible outside the Hubble radius.

The outline of the analysis is as follows: we first study the evolution of the quantum field to determine the deviations of the equation of state from an exact de Sitter equation. In terms of $\phi_0(t)$, w and c_s^2 are

$$w(t) = -1 + \frac{\dot{\phi}_0^2(t)}{\rho} \tag{29}$$

$$c_s^2(t) = -1 - \frac{2\ddot{\phi}_0(t)}{3H\dot{\phi}_0(t)} . \tag{30}$$

In a second step we use the resulting expressions to determine the growth in Φ_H via Eq. (28).

Consider first the evolution of the classical field $\phi_0(t)$ of Eq. (21). During a first period $t_i < t < t_B$ the field remains close to the top of the potential of Fig. 2, i.e. $\phi_0(t) \sim H$, $\dot{\phi}_0(t) \sim H^2$. Hence by Eqs. (29) and (30) $w(t) \sim -1$ and $c_s^2(t) \sim -1$. The length τ of this period, the main inflationary period in new inflationary universe models, must exceed $\sim 50\ H^{-1}$ in order to solve the horizon and flatness problems[1)] of the universe. In the second period of the de Sitter phase, for $t_B < t < t_R$. $\phi_0(t)$ rapidly increases; it begins to "roll down the potential hill" towards the minimum at $\phi = \sigma$. $\dot{\phi}_0(t)$ increases from a value of order H^2 to a value of order σ^2. Hence $w(t)$ approaches zero while c_s^2 becomes very negative, approaching a value of the order $- \sigma/H$. The equation of state as a function of time is sketched in Fig. 3. During and after reheating both $w(t)$ and $c_s^2(t)$ rapidly approach their FRW values $w = c_s^2 = 1/3$.

The implications for the evolution of Φ_H are not hard to analyze qualitatively. In the first period $w(t) = c_s^2(t)$. Hence by Eq. (28) [$I(t)$ is negligible!] one mode will be constant in time while the second decays exponentially. Thus Φ_H remains constant. In the second period $|c_s^2|$ increases by many orders of magnitude. $\dot{\Phi}_H$ initially vanishes, so to compensate in Eq. (28) $\ddot{\Phi}_H$ must become large and positive. In terms of physics: the rapid change in the equation of state induces an increase in Φ_H. This is sketched in Fig. 4. In the FRW period $w(t)$ again equals $c_s^2(t)$ and thus by the same argument as before Φ_H will remain constant.

Energy density perturbations will thus be much larger when they reenter the Hubble radius at $t_f(k)$ than they were when they left at $t_i(k)$. To obtain the amplification factor and thus the final value of $\delta\rho/\rho$ we need a quantitative analysis.

The simplest way to proceed is to reformulate the equation of motion (28) as an approximate conservation law. We define a variable ξ by[5)]

$$\xi = \frac{2}{3}\,\frac{\Phi_H + H^{-1}\dot{\Phi}_H}{1+w} + \Phi_H\left(1 + \frac{2}{9}\left(\frac{k}{aH}\right)^2 \frac{1}{1+w}\right). \tag{31}$$

It is not hard to show that, up to terms suppressed by an additional factor $(k\,a^{-1}\,H^{-1})^2$ outside the Hubble radius, $\dot{\xi} = 0$ is equivalent to the homogeneous equation of motion for Φ_H. Hence Φ_H at $t_f(k)$ can be easily obtained by equating ξ at $t_i(k)$ and $t_f(k)$. Since $\dot{\Phi}_H(t_f)$ vanishes and since $H^{-1}\dot{\Phi}_H(t_i)$ equals $\Phi_H(t_i)$ up to factors of order unity, we obtain

$$\Phi_H(t_f) = O(1)(1 + w(t_i))^{-1}\,\Phi_H(t_i) \tag{32}$$

or equivalently

$$\frac{\delta\rho}{\rho}(t_f) = O(1)(1 + w(t_i))^{-1}\,\frac{\delta\rho}{\rho}(t_i)\ . \tag{33}$$

The amplification factor for the amplitude of energy density perturbations is $(1 + w(t_i))^{-1}$. The first important conclusion is that the final energy density fluctuation amplitude will be independent of the equation of state between the times when scales of cosmological interest leave and reenter the Hubble radius. In particular, additional phase transitions will have no effect. Nor will the details of reheating be important. Since in most specific models the equation of state is essentially constant during the time interval when scales of cosmological interest leave the Hubble radius, the amplification factor will be scale independent. This confirms the second of our heuristic arguments in the introduction: a scale invariant spectrum at t_i will produce a scale invariant spectrum at t_f.

In the new inflationary universe[2)] by Eq. (29)

$$1 + w(t_i) = O(1)\left(\frac{H}{\sigma}\right)^4\ . \tag{34}$$

Hence by Eqs. (23) and (33) the final amplitude of $\delta\rho/\rho$ is of order unity. This is the famous "fluctuation problem"[11]: the amplitude exceeds the value required for models of galaxy formation by four or five orders of magnitude.

To apply our analysis of the classical evolution of fluctuations to general particle physics models we must reexpress $\Phi_H(t_i)$ as well as the amplification factor in terms of particle physics parameters.[5] Evaluating $\delta\rho/\rho$ in comoving coordinates we obtain

$$\Phi_H(t_i) = O(1)\,\{\dot{\phi}_0\,\delta\dot{\phi}_0 - \ddot{\phi}_0\,\delta\phi_0\}\,(t_i)\,\rho^{-1} \; . \tag{35}$$

In many inflationary universe models the "slow rolling" approximation is valid, i.e. $\ddot{\phi}_0$ is negligible for $t < t_i$. Using the classical equation of motion for $\phi_0(t)$

$$\ddot{\phi}_0 + 3H\dot{\phi}_0 = - V'(\phi_0) \tag{36}$$

we can combine Eqs. (32) and (35) to give

$$\frac{\delta\rho}{\rho}\,(t_f) = O(1)\,\frac{V''(\phi_0)}{\dot{\phi}_0 H}\,\delta\phi \; . \tag{37}$$

In examples for which $\ddot{\phi}_0 > 3H\dot{\phi}_0$ we similarly obtain

$$\frac{\delta\rho}{\rho}\,(t_f) = O(1)\,\frac{V'(\phi_0)}{\dot{\phi}_0^{\,2}}\,\delta\phi \; . \tag{38}$$

The quantities on the right-hand side of Eqs. (37) and (38) must be evaluated at t_i.

In most concrete particle physics models a single scale, namely H, determines all the characteristics of the potential for ϕ near $\phi(t_i)$. In this case we recover the "magic" formula of Ref. 11:

$$\frac{\delta\rho}{\rho}\,(t_f) = O(1)\,H\,\frac{\delta\phi}{\dot{\phi}_0}\,(t_i) \; . \tag{39}$$

In order to solve the fluctuation problem, i.e. to obtain a sufficiently small value for $\delta\rho/\rho(t_f)$ we must change the shape of the scalar field potential $V(\phi)$ such that $\dot{\phi}_0(t_i)$ increases and/or $V''(\phi_0(t_i))$ decreases. The first requires that the slope of the potential be large (compared to the value in the simplest new inflationary universe model), the second that the curvature be artificially small. We also need a very small curvature to obtain enough inflation.

Models which satisfy these criteria have recently been proposed.[19) They involve either two different physical scales put in by hand (and thus lead to a hierarchy problem), or else they involve fine tuning free parameters. It is, however, premature to focus on the fine tuning problems for specific particle physics models when so many fundamental questions, e.g. the origin of the cosmological constant, remain unanswered. This is why the focus of this lecture was on the basic mechanism by which inhomogeneities are generated and propagated until they enter the Hubble radius.

ACKNOWLEDGMENTS

Foremost I thank Ron Kahn for the fun of our collaboration. For useful discussions I am grateful to Jim Bardeen, Doug Eardley, Bill Press, Lenny Susskind, Nick Tsamis, Michael Turner, and Tony Zee. This material is based upon research supported in part by the National Science Foundation under Grant No. PHY77-27084, supplemented by funds from the National Aeronautics and Space Administration.

REFERENCES

1. A. Guth, Phys. Rev. D 23, 347 (1981).
2. A. Linde, Phys. Lett. 108 B, 389 (1982); A. Albrecht and P. Steinhardt, Phys. Rev. Lett. 48, 1220 (1982).
3. E. Harrison, Phys. Rev. D 1, 2726 (1970); Ya. Zeldovich, Mon. Not. R. astr. Soc. 160, 1p (1972).
4. W. Press, Physica Scripta 21, 702 (1980).
5. R. Brandenberger and R. Kahn, Phys. Rev. D 29, 2172 (1984).
6. S. Hawking, Comm. Math. Phys. 43, 199 (1975).
7. G. Gibbons and S. Hawking, Phys. Rev. D 15, 2738 (1977); A. Lapedes, J. Math. Phys. 19, 2289 (1978).
8. R. Brandenberger and R. Kahn, Phys. Lett. 119 B, 75 (1982).
9. T. Bunch and P. Davies, Proc. R. Soc. London A 360, 117 (1978).
10. R. Brandenberger, Phys. Lett. 129 B, 397 (1983).
11. J. Bardeen, P. Steinhardt, and M. Turner, Phys. Rev. D 28, 679 (1983); A. Guth and S.-Y. Pi, Phys. Rev. Lett. 49, 1110 (1982); S. Hawking, Phys. Lett. 115 B, 295 (1982); A. Starobinsky, Phys. Lett. 117 B, 175 (1982).
12. R. Brandenberger, ITP preprint NSF-ITP-84-03, submitted to Nucl. Phys. B.
13. Similar ideas have been discussed by J. Bardeen, unpublished (1983).
14. A. Vilenkin and L. Ford, Phys. Rev. D 26, 1231 (1982).
15. A. Linde, Phys. Lett. 116 B, 335 (1982).
16. J. Peebles, "The large scale structure of the universe" (Princeton University Press, Princeton, 1980).
17. J. Bardeen, Phys. Rev. D 22, 1882 (1980).
18. R. Brandenberger, R. Kahn, and W. Press, Phys. Rev. D 28, 1809 (1983).

19. J. Ellis, D. Nanopoulos, K. Olive, and K. Tamvakis, Phys. Lett. 120 B, 331 (1983); D. Nanopoulos, K. Olive, M. Srednicki, and K. Tamvakis, Phys. Lett. 123 B, 41 (1983); A. Linde, Phys. Lett. 132 B, 317 (1983); A. Albrecht *et al.*, Nucl. Phys. B 229, 528 (1983); P. Steinhardt and M. Turner, Fermilab preprint 84/19-A (1984); Q. Shafi and A. Vilenkin, Phys. Rev. Lett. 52, 691 (1984); S.-Y. Pi, Phys. Rev. Lett. 52, 1725 (1984); S. Gupta and H. Quinn, SLAC preprint 3269 (1983).

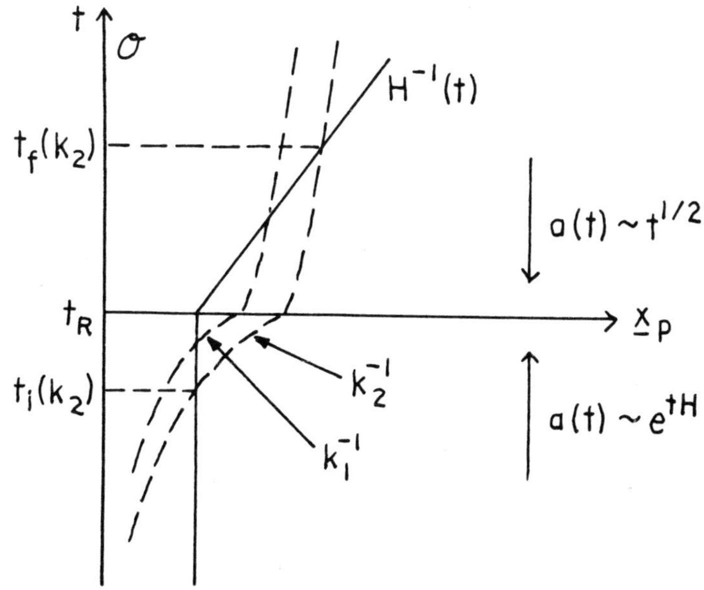

Figure 1: The cosmology of inflationary universe models.
Sketch of the evolution of fluctuations on two scales in physical coordinates x_p.

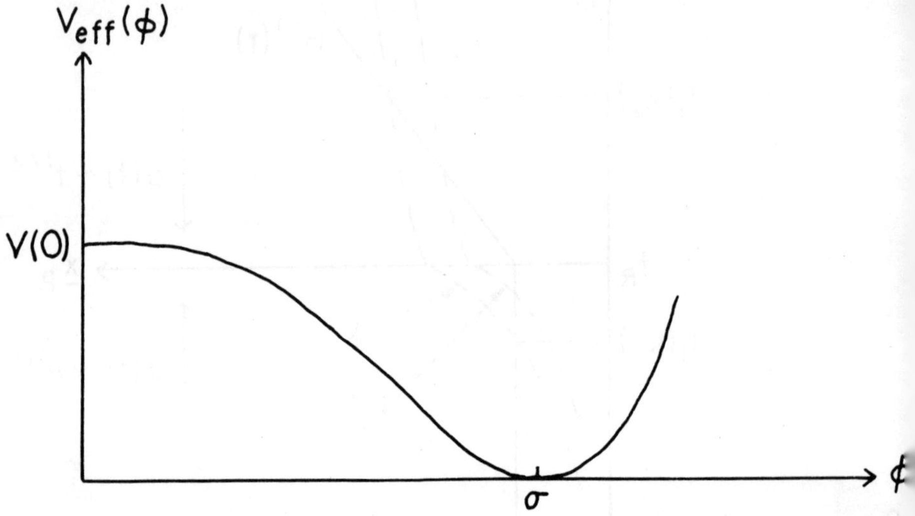

Figure 2: Sketch of the effective potential in the new inflationary universe.

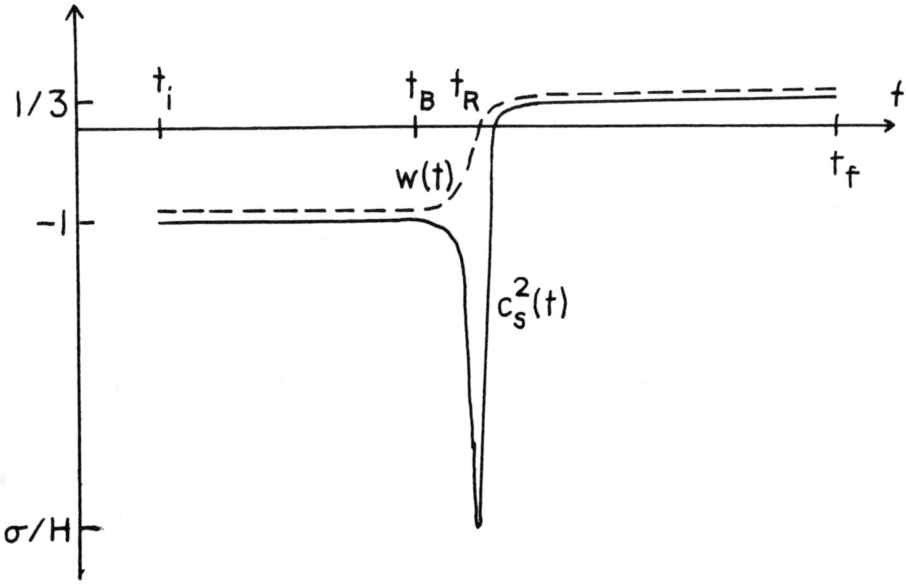

Figure 3: Equation of state in the new inflationary universe (below -1 the scale is logarithmic).

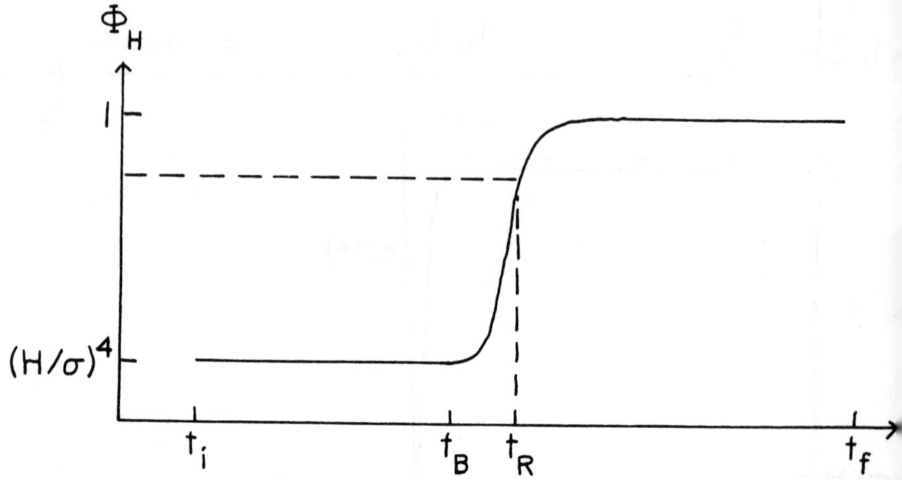

Figure 4: Growth of Φ_H in the new inflationary universe (the vertical scale is logarithmic).

GALAXY AND CLUSTER FORMATION IN A UNIVERSE DOMINATED BY COLD DARK MATTER

Joel R. Primack

Stanford Linear Accelerator Center

Stanford University, Stanford, California 94305

and

Santa Cruz Institute for Particle Physics

University of California, Santa Cruz, California 95064

The dark matter (DM) that appears to be gravitationally dominant on all astronomical scales larger than the cores of galaxies[1] can be classified, on the basis of its characteristic free-streaming damping mass M_D, as hot ($M_D \sim 10^{15} M_\odot$), warm ($M_D \sim 10^{11} M_\odot$), or cold ($M_D < 10^8 M_\odot$). For the case of cold DM, the shape of the DM fluctuation spectrum is determined by (a) the primordial spectrum (on scales larger than the horizon), which is usually assumed to have a power spectrum of the form $|\delta_k|^2 \propto k^n$ (inflationary models[3] predict the "Zeldovich spectrum" $n = 1$); and (b) "stagspansion",[2] the stagnation of the growth of DM fluctuations that enter the horizon while the universe is still radiation-dominated, which flattens the fluctuation spectrum for $M \lesssim 10^{15} M_\odot$.[4-6]

An attractive feature of the cold dark matter hypothesis is its considerable predictive power: the post-recombination fluctuation spectrum is calculable, and it in turn governs the formation of galaxies and clusters. Good agreement with the data is obtained for a Zeldovich spectrum of primordial fluctuations.

1. WHY COLD DM?

There are strong astrophysical arguments that the DM does not consist of any form of ordinary matter ("baryons").[2] Although these arguments are not entirely compelling, they are sufficiently convincing to have motivated both astrophysicists and particle physicists to consider seriously the possibility that the DM consists of some other sort of matter.

If a species of neutrino is the gravitationally dominant component of the universe, [8,9] its mass $m_\nu = 100\ \Omega\ h^2 eV$ (where $h = H_0/100\ km\, s^{-1} Mpc^{-1}$ lies in the range $1/2 \leq h \leq 1$) implies a free-streaming damping mass $M_D \sim 10^{15} M_\odot$ corresponding to hot DM. Since fluctuations of galactic mass $\sim 10^{8-12} M_\odot$, much smaller than M_D, are strongly damped, galaxies can only form in a neutrino-dominated universe after fluctuations of supercluster mass $\sim 10^{15} M_\odot$ have collapsed. Partly because this type of DM has been the most intensively studied, a number of potential problems have been identified — for example, the late formation of supercluster "pancakes", at $z_p \lesssim 2$,[10] which subsequently fragment into galaxies. However, the best limits on galaxy ages coming from globular clusters and other stellar populations, plus the possible association of QSO's with galactic nuclei, indicate that galaxy formation took place before $z = 3$.[11] This is inconsistent with the "top-down" neutrino theory, in which superclusters form before galaxies rather than after them.

Another problem with the neutrino picture is that large clusters of galaxies can accrete neutrinos more efficiently than ordinary galactic halos, which have lower escape velocities. One-dimensional numerical simulations predict that the ratio of total to baryonic mass M/M_b should be ~ 5 times larger for clusters ($\sim 10^{14} M_\odot$) than for ordinary galaxies ($M \sim 10^{12} M_\odot$).[12] While there is evidence that the mass-to-light ratio M/L does increase with scale, there is also considerable evidence that the more physically meaningful ratio of total to luminous mass M/M_{lum} remains constant from large clusters through groups of galaxies, binary galaxies, and ordinary spirals. (M_{lum}, which is the mass visible in galactic stars and gas plus hot, X-ray emitting gas, is $\leq M_b$, since an unknown fraction of the baryons is invisible—*e.g.*, in the form of diffuse ionized intergalactic gas at $T \sim 10^4$ K.)

This is illustrated in Fig. 1, which presents the available data for M/L and M/M_{lum}. The fact that the total-to-luminous mass of rich clusters is similar to that of galaxies including their massive halos, even though the clusters' mass-to-light ratio is larger, is due mainly to the different stellar population in the ellipticals, and the large contribution of X-ray emitting gas to M_{lum}, in rich clusters. (In very rich clusters such as Coma, there is $\sim 2-5$ times as much mass in hot gas as there is in stars.)

Finally, preliminary velocity dispersion data for Draco, Carina, and Ursa Minor as well as theoretical arguments[15] suggest that a significant amount of DM may reside in dwarf spheroidal galaxies. Because of the low velocity dispersion of dwarf galaxies, phase space constraints give a lower limit of $m >$ 500 eV for the mass of particles comprising this DM.[16] The present velocity dispersion estimates are uncertain owing to possible stellar oscillations, mass outflow, and binary motions, but these effects can be discovered and eliminated with careful monitoring. The mass limit of 500 eV would rule out neutrinos as the halo DM in dwarf spheroidal galaxies. If we assume that the DM has essentially the same composition everywhere, as is suggested by the constancy of M/M_{lum} in Fig. 1, then the DM is not neutrinos.[17]

The DM in the dwarf spheroidal halos is probably not warm DM either. Warm DM first collapses on a scale $\sim 10^{11} M_\odot$ with velocity dispersion $\sigma \sim$

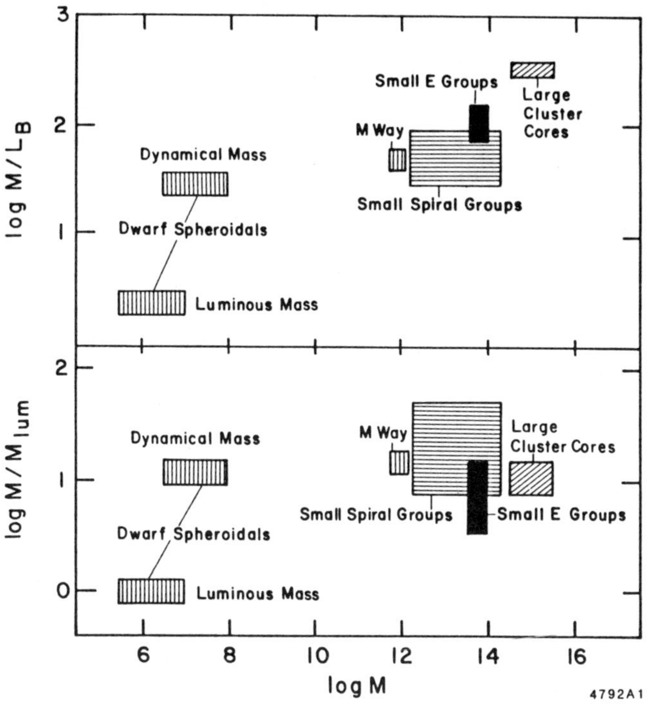

FIGURE 1. Mass-to-light ratio, M/L_B, and total-to-luminous mass, M/M_{lum}, for structures of various size in the universe. Although M/L_B increases systematically with mass, the more physically meaningful ratio M/M_{lum} appears to be constant on all scales within the errors.[13]

$10^2 km/s$, and too little could be captured by dwarf spheroidals, having $\sigma \sim 10\, km/s$, to form the heavy halos indicated by the observations.

Besides the evidence just summarized against hot and warm DM, a further reason to consider cold DM is the existence of several plausible physical candidates, including axions of mass $\sim 10^{-5}$ eV;[21,22] heavy stable particles, such as the photino, with a mass $\gtrsim 0.5$ GeV and very weak interactions;[23] and primordial black holes[24] with $10^{17} g \gtrsim m_{PBH} \gtrsim M_\odot$.[7] Still another exotic cold DM candidate has recently been proposed by Witten: "nuggets" of $u - s - d$ symmetric quark matter.[25] There is thus no shortage of cold DM candidate particles — although there is admittedly no direct evidence that any of them actually exists.

2. THE COLD DM FLUCTUATION SPECTRUM

We will follow the current conventional wisdom and assume that the primordial fluctuations were adiabatic. In the standard formulation, fluctuations $\delta \equiv \delta\rho/\rho$ grow as $\delta \sim a^2$ on scales larger than the horizon, where $a = (1+z)^{-1}$ is the scale factor normalized to 1 at the present. When a fluctuation enters the horizon in the radiation-dominated era, the photons (together with the charged particles) oscillate as an acoustic wave, and the non-interacting neutrinos freely stream away (they are still relativistic, since in the cold DM case their masses are $\ll 30$ eV). As a result, the main driving terms for the growth of δ_{DM} disappear and the growth accordingly stagnates ("stagspansion") until matter dominates; see Fig. 2. Matter domination first occurs at $z = z_{eq}$, where

$$\begin{aligned} z_{eq} &= 4.2 \times 10^4 h^2\ \Omega\ (1 + 0.68 N_\nu)^{-1} \\ &= 2.5 \times 10^4 h^2\ \Omega \text{ for } N_\nu = 3 \end{aligned} \tag{1}$$

The first study of the growth of cold DM fluctuations was the numerical calculations of Peebles,[4] who for simplicity ignored neutrinos: $N_\nu = 0$ in (1). Subsequent numerical calculations have included the effects of the known neutrino species ($N_\nu = 3$, $m_\nu \approx 0$) both outside and inside the horizon.[2,5,6,26,27,28] Numerically, the largest effect of including neutrinos is the change in z_{eq}.

It is instructive to make the further approximation of setting $\delta_{\gamma+b} = \delta_\nu = 0$ once a fluctuation is inside the horizon. Then one can analytically match the solution for $a > a_{\text{horizon}}$

$$\delta_{DM}(a) = A_1 D_1(a) + A_2 D_2(a)\ , \tag{2}$$

$$D_1 = 1 + 1.5y \quad \text{where} \quad y = a/a_{eq}\ , \tag{3a}$$

$$D_2 = D_1 \ell n \left[\frac{(1+y)^{1/2} + 1}{(1+y)^{1/2} - 1} \right] - 3(1+y)^{1/2}\ . \tag{3b}$$

to the growing mode $\delta_{DM} \sim a^2$ for $a < a_{\text{horizon}}$. Matching the derivatives requires $A_2 D_2$ comparable to $A_1 D_1$ but opposite in sign. For $a \gg a_{\text{horizon}}$ only the growing solution D_1 survives, which explains the moderate growth in δ_{DM} between horizon crossing and matter dominance. In the limit of large k, one finds $\delta_k \propto k^{n/2-2} \ell n\, k$. Correspondingly, for $M \ll M_{eq} \approx 10^{16} M_\odot$, the rms fluctuation in the mass within a random sphere containing average mass M is $\delta M/M \propto |\ell n\, M|^{3/2}$. Some authors have considered only the Meszaros solution (3a) and erroneously inferred that the fluctuation spectrum would be essentially flat for $M < M_{eq}$ for a Zeldovich primordial spectrum, which would then be inconsistent with observations.[29]

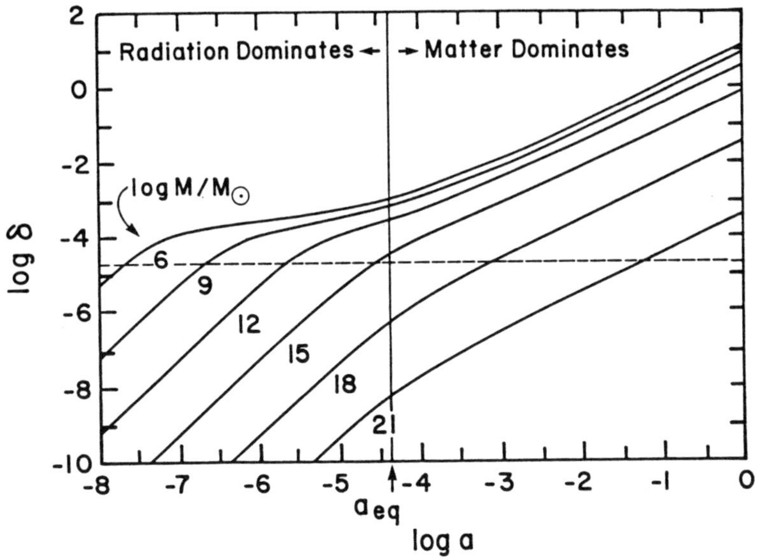

FIGURE 2. Numerical results for the growth of $\delta = k^{3/2}\delta_k$ versus scale factor a for fluctuations of various masses $M = \frac{4}{3}\pi^4 k^{-3}\rho_c$. The curves are drawn for $n = 1$, $\Omega = h = 1$, and a baryonic to total mass ratio of 0.1. The vertical line represents the value of a when the universe becomes matter dominated, and the dashed line shows the (constant for $n = 1$) value of δ when each mass scale crosses the horizon. These curves illustrate the stagnation of perturbation growth after small mass scales cross the horizon and show why at late times $\delta(k)$ is nearly flat for large k (small M). (From Ref. 5.)

Our numerical results for $\delta M/M$ are shown in Fig. 3 for $\Omega = h = 1$, assuming a Zeldovich ($n = 1$) spectrum (reflected in $\delta M/M \propto M^{-2/3}$ for $M > M_{eq}$). For either h or Ω less than unity, $\delta M/M$ is somewhat flatter.[7,6]

3. GALAXY AND CLUSTER FORMATION

The key features of galaxy formation in the cold DM picture are these: after recombination (at $z_{rec} \approx 1300$) the amplitude of the baryonic fluctuations rapidly grows to match that of the DM fluctuations; smaller-mass fluctuations grow to nonlinearity and virialize, and then are hierarchically clustered within successively larger bound systems; and finally the ordinary matter in bound systems of total mass $\sim 10^{8-12} M_\odot$ cools rapidly enough within their DM halos

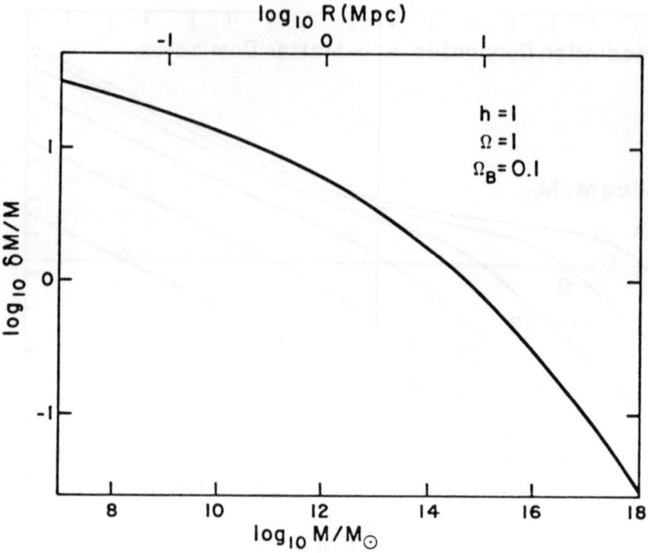

FIGURE 3. The rms mass fluctuations within a randomly placed sphere of radius R in a cold DM universe. The curve is normalized[4] at 8 Mpc and assumes a primordial Zeldovich ($n = 1$) fluctuation spectrum, and $h = \Omega = 1$. (From Ref. 2.)

to form galaxies, while larger mass fluctuations form clusters.

At any mass scale M, when the fluctuation $\delta M/M$ approaches unity, nonlinear gravitational effects become important. The fluctuation then separates from the Hubble expansion, reaches a maximum radius, and begins to contract. Spherically symmetric fluctuations, for example, contract to about half their maximum radii. During this contraction, violent relaxation[30] due to the rapidly varying gravitational field converts enough potential energy into kinetic energy for the virial theorem, $\langle PE \rangle = -2 \langle KE \rangle$, to be satisfied. After virialization, the mean density within a fluctuation is roughly eight times the density corresponding to the maximum radius of expansion.[31]

Since the cold-DM fluctuation spectrum $\delta M/M$ is a decreasing function of M, smaller mass fluctuations will, on the average, become nonlinear and begin to collapse at earlier times than larger mass fluctuations. Small mass bound systems are subsequently clustered within larger mass systems, which go nonlinear at a later time. This hierarchical clustering of smaller systems into larger and yet larger gravitationally bound systems begins at the baryon Jeans mass ($M_{J,b} \sim 10^6 M_\odot$ at recombination) and continues until the present time. The baryonic substructures within larger mass clusters will be disrupted by subsequent virialization of the clusters unless significant mass segregation between baryons and DM has occured prior to cluster virialization. Hence, in order to maintain their existence as a separate substructure, the baryons must cool and gravitationally condense within their massive DM halos *before* virialization occurs on larger scales.[32]

Figure 4 shows the density of ordinary (baryonic) matter versus internal kinetic energy (temperature) of typical fluctuations of various sizes, just after virialization, calculated from $\delta M/M$ of Fig. 3. This is superimposed upon the Rees-Ostriker[33] cooling curves (for which cooling time equals gravitational free fall time) and data on galaxies (with kinetic energy determined from rotation velocity for spirals and velocity dispersion for ellipticals).[34]

Fluctuations that start with greater amplitude than average will turn around earlier, at higher density, and thus lie below the virialization curve on Fig. 4. As the baryons in a virialized fluctuation dissipate, their density will initially increase at constant T within the surrounding isothermal halo of dissipationless material (DM), and then T will increase as well when the baryon density exceeds the DM density, as suggested by the dashed line in the figure. The Zeldovich primordial spectrum is more consistent with the data on Fig. 4 than an $n = 2$ (or $n = 0$) primordial spectrum, which lies too low (too high) on the figure compared to the galaxies. With the Zeldovich spectrum, the important conclusion is that one should observe dissipated systems with large halos having total mass $10^8 M_\odot \lesssim M \lesssim 10^{12} M_\odot$. This is essentially the range of observed galaxy masses.

While the $n_b - T$ diagram (Fig. 4) is useful for comparing data and predictions with the cooling curves, it is also useful to consider total mass M versus T, as in Fig. 5. This avoids having to take into account the differing amounts of baryonic dissipation suffered by various galaxies. The heavy solid and dashed curves again correspond to the $n = 1$ cold DM spectrum, for ($\Omega = 1, h = 0.5$) and ($\Omega = 0.2, h = 1$) respectively. It is striking that the galaxies in the $M - T$ diagram lie along lines of roughly the same slope as these curves. This occurs because the effective slope of the $n = 1$ cold DM fluctuation spectrum in the galaxy mass range is $n_{eff} \approx -2$, which corresponds to the empirical Tully-Fisher and Faber-Jackson laws: $M \propto v^4$. The light dashed lines in Fig. 5 are the post-virialization curves for primordial fluctuation spectra with $n = 0$ (white noise) and $n = 2$. Again, the $n = 1$ (Zeldovich) spectrum is evidently the one that is most consistent with the data.

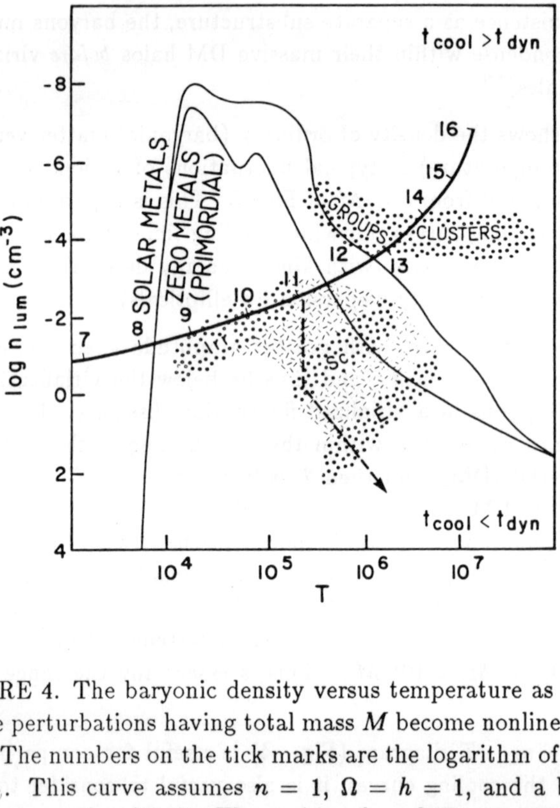

FIGURE 4. The baryonic density versus temperature as root-mean-square perturbations having total mass M become nonlinear and virialize. The numbers on the tick marks are the logarithm of M in units of $M_\odot$. This curve assumes $n = 1$, $\Omega = h = 1$, and a baryonic to total mass ratio of 0.07. The region where baryons can cool within a dynamical time lies below the cooling curves. Also shown are the positions of observed galaxies, groups, and clusters of galaxies. The dashed line represents a possible evolutionary path for dissipating baryons. (From Ref. 2.)

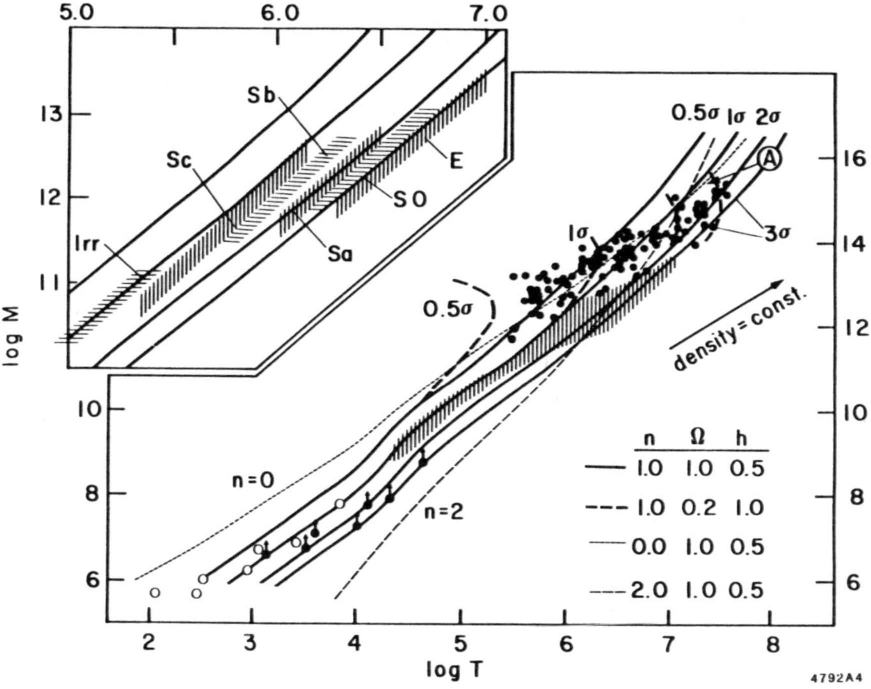

Figure 5. Total mass M versus virial temperature T. The quantity T is $\mu V^2/3k$, where μ is mean molecular weight (≈ 0.6 for ionized, primordial H + He) and k is Boltzmann's constant. M for groups and clusters is total dynamical mass. For galaxies, M is assumed to be 10 M_{lum} (corresponding to Fig. 1). If dwarf spheroidals actually have $M/L_B = 30$, they may have suffered baryon stripping[16], in which case M is a lower limit (arrows). Details of the region occupied by massive galaxies are shown in the inset in upper left.

Model curves represent the equilibria of structures that collapse dissipationlessly from the cold dark matter initial fluctuation spectra with $n = 1$. The curves labeled 1σ refer to fluctuations with $\delta M/M$ equal to the rms value. Curves labeled 0.5σ, 2σ, and 3σ refer to fluctuations having 0.5, 2, and 3 times the rms value. Heavy curves: $\Omega = 1$, $h = 0.5$; dashed curves: $\Omega = 0.2$, $h = 1$; these cases were chosen to span the astrophysically interesting range. In addition to the n = 1 curves, two 1σ curves for $n = 0$ and $n = 2$ are also shown (light dashes).

Major conclusions from the figure: 1) Either set of curves for $n = 1$ (Zeldovich spectrum) provides a good fit to the observations over 9 orders of magnitude in mass. Curves with $n = 0$ and $n = 2$ do not fit as well. 2) The apparent gap between galaxies and groups and clusters in Fig. 4 (which stems from baryonic dissipation) vanishes in this figure, and the clustering hierarchy is smooth and unbroken from the smallest structures to the largest ones. 3) The Fisher-Tully and Faber-Jackson laws for galaxies ($M \propto V^4$ or T^2) arise naturally as a consequence of the slope of the cold DM fluctuation spectrum in the mass region of galaxies. 4) Groups and clusters are distributed around the $n = 1$ loci about as expected. The apparent upward trend among the groups is not physically meaningful but arises from their selection as minimum-density enhancements (see constant-density arrow). 5) The exact locations of galaxies are somewhat uncertain. In particular, the temperatures of E's and S0's may be overestimated owing to the use of nuclear rather than global velocity dispersions. Taken at face value, however, the data suggest that early-type galaxies (E's and S0's) arise from high-$\delta M/M$ fluctuations, whereas late-type galaxies (Sc's and Irr's) arise from low-$\delta M/M$ fluctuations. 6) Groups and clusters appear to fill a wider band than galaxies. If real, this difference may indicate that very weak, low-$\delta M/M$ fluctuations on the mass scale of galaxies once existed but did not give rise to visible galaxies. This suggests further that galaxy formation, at least in some regions of the universe, may not have been fully complete and that galaxies are therefore not a reliable tracer of total mass. 7) There seems to be a real trend along the Hubble sequence to increasing mass among early-type galaxies. Neither this trend nor the rather sharp demarcation between galaxies and groups and clusters is fully understood. (This figure is from Ref. 7.)

The points in Fig. 5 represent essentially all of the clusters identified by Geller and Huchra[35] in the CfA catalog within 5000 km s^{-1}. The cluster data lie about where they should on the diagram, and even the statistics of the distribution seem roughly to correspond to the expectations represented by the $0.5, 1, 2$, and 3σ curves.

Notice that spiral galaxies lie roughly along the 1σ curve while elliptical galaxies lie along the 2σ curve. Although this displacement is not large compared to the uncertainties, it is consistent with the fact that more than half of all galaxies are spirals, while only about 15 percent are ellipticals. In hierarchical clustering scenarios, it seems likely that the higher σ fluctuations will develop rather smaller angular momenta, as measured by the dimensionless parameter λ

($= JE^{\frac{1}{2}}G^{-1}M^{-\frac{5}{2}}$). There are two reasons for this: high-overdensity fluctuations collapse earlier than average fluctuations, and are thus typically surrounded by a relatively homogeneous matter distribution;[36] also, higher amplitude fluctuations are typically rounder[37] and consequently have lower quadrupole moments. Both effects result in less torque. This difference appears to exist with either white noise or a flatter spectrum, but to be somewhat larger in the latter case. If high σ fluctuations have little angular momentum, their baryons can collapse by a large factor in radius, forming high-density ellipticals and spheroidal bulges, as shown in Fig. 4. Since, with a flat spectrum, higher σ fluctuations occur preferentially in denser regions destined to become rich clusters, one expects[38] to find more ellipicals there—as is observed.[39] Note that the rich clusters lie along the same 2 and 3σ curves in Fig. 5 as do the elliptical galaxies.

Note also that while the galaxy data lies below the rms virialization curve, the data on groups and clusters of galaxies lies more or less evenly around it. This suggests that galaxy formation may be an inefficient process, with lower-amplitude fluctuations of galaxy mass not giving rise to visible galaxies.[40]

Presumably the collapse of the low-λ protoellipical galaxies is halted by star formation well before a flattened disk can form, yielding a stellar system of spheroidal shape. The mechanism governing the onset of star formation in these systems is unfortunately not yet understood, but may involve a threshold effect which sets in when the baryon density exceeds the DM halo density by a sufficient factor.[32,41] Disks (spirals and irregulars) form from average, higher-λ protogalaxies, which, for a given mass, are larger and more diffuse than their protoelliptical counterparts. The collapse of disks thus occurs via relatively slow infall of baryons from $\sim 10^2$ kpc, halted by angular momentum. Infall from such distances is consistent both with the extent of dark halos inferred from observations and with the high angular momenta of present-day disks ($\lambda \sim 0.4$).[42] The location of the galaxies in Fig. 4 is consistent with these ideas if the baryons in all galaxies collapsed by roughly the same factor, about an order of magnitude, but somewhat less for late-type irregulars and somewhat more for early-type E's and spheroidal bulges.

It has been theorized that the Hubble sequence originates in the distribution of either the initial angular momenta or else the initial densities[44] of protogalaxies. However, if overdensity and angular momentum are linked, with the high-σ fluctuations having lower λ, then these two apparently competitive theories become the opposite sides of the same coin.

It is interesting to ask whether the cold DM picture can account for the wide range of morphologies displayed by clusters of galaxies in X-ray[45] and optical-band[46] observations, ranging from regular, apparently relaxed configurations to complex, multicomponent structures. Preliminary results are encouraging. In particular, simulations show that large central condensations form quickly and can grow by subsequent mergers to form cD galaxies if most of the DM is in halos around the baryonic substructures, as expected for cold DM, but not if the DM is distributed diffusely.[47]

Consider finally the difference in Fig. 5 between the solid and dashed lines. The dashed lines, representing a lower-density universe ($\Omega = 0.2$), curve backward at the largest masses and lie far away from the circle representing the cores of the richest clusters, Abell classes 2 and 3. Since these regions of very high galaxy density contain at least several percent of the mass in the universe, the circle should lie between the 2 and 3σ lines (assuming Gaussian statistics). It does so for the solid ($\Omega = 1$) lines, but not for the dashed lines. At face value, this is evidence favoring an Einstein-de Sitter universe for cold DM. However, there are at least two reasons why this argument should probably not be taken too seriously. First, the velocity dispersions represented by the Abell cluster circle in Fig. 5 correspond to the cluster cores. The model curves on the other hand refer to the entire virialized cluster, over which the velocity dispersion is considerably lower (as indicated by the arrow attached to the circle in Fig. 5). Second, the assumption of spherical symmetry used in obtaining both sets of curves in the figure is only an approximation. The initial collapse is probably often quite anisotropic—more like a Zeldovich pancake than a sphere. It is therefore preferable to compare these data with N-body simulations rather than with the simple model represented by the curves in Fig. 5. This will require N-body simulations of large dynamical range, which can perhaps be achieved by putting many mass points into one cell of the P^3M−type simulations.[48] Until this becomes possible the data in the figure do not allow a clear-cut discrimination between the $\Omega = 0.2$ and $\Omega = 1$ cases, especially if the Hubble parameter h is allowed to vary simultaneously within the observationally allowed range, as has been assumed.

Other data are also relevant to the determination of Ω, of course.[7] For example, the latest observations of small-angle fluctuations in the cosmic background radiation[49] imply[26,27] $\Omega \geq 0.2h^{-4/3}$, unless there is significant reheating of the intergalactic medium after recombination.[50]

4. REMARKS

A universe with ~ 10 times as much cold dark matter as baryonic matter provides a remarkably good fit to the observed universe. This model predicts roughly the observed mass range of galaxies, the dissipational nature of galaxy collapse, and the observed Faber-Jackson and Tully-Fisher relations. It also gives dissipationless galactic halos and clusters. In addition, it may also provide natural explanations for galaxy-environment correlations and for the differences in angular momenta between ellipticals and spiral galaxies. Finally, the cold DM picture seems reasonably consistent with the observed large-scale clustering, including superclusters and voids.[51] In short, it appears to be the best model presently available and merits close scrutiny and testing in the future.

Acknowledgments

This paper is based on research done in collaboration with George Blumenthal, Sandra Faber, and Martin Rees[52] — all of whom I would like to thank for their efforts to teach me astrophysics! I am grateful to Sidney Drell for the hospitality of SLAC, and I acknowledge partial support from the National Science Foundation under grant PHY-81-15541.

REFERENCES

1. Faber, S. M. and Gallagher, J. S. *Ann. Rev. Astron. Astrophys.* **17**, 135-187 (1979).
2. Primack, J. R. and Blumenthal, G. R. *Formation and Evolution of Galaxies and Large Structures in the Universe* (eds. J. Audouze and J. Tran Thanh Van) 163-183 (Reidel, Dordrecht, Holland, 1983). Primack, J. R. and Blumenthal, G. R. *Fourth Workshop on Grand Unification* (eds. Weldon, H. A., Langacker, P. and Steinhardt, P. J.) 256-288 (Birkhauser, Boston, 1983).
3. Brandenberger, R. (these Proceedings).
4. Peebles, P. J. E. *Astrophys. J. Lett.* **263**, L1-L5 (1982).
5. Primack, J. R. and Blumenthal, G. R. *Clusters and Groups of Galaxies* (eds. Mardirossian, F., Giuricin, G., and Mezzetti, M., Reidel, Dordrecht, Holland, in the press).
6. Blumenthal, G. R. and Primack, J. R., in prep. (1984).
7. Blumenthal, G. R., Faber, S. M., Primack, J. R. and Rees, M. J. *Nature* (in the press).
8. Zeldovich, Ya. B., Einasto, J. and Shandarin, S. F. *Nature* **300**, 407-413 (1982).
9. Shandarin, S. F., Doroshkevich, A. G. and Zeldovich, Ya. B. *Sov. Phys. Usp.* **26**, 46-76 (1983).
10. Frenk, C., White, S. D. M. and Davis, M. *Astrophys. J.* **271**, 417-430 (1983). Kaiser, N. *Astrophys. J.* **273**, L17-L20 (1983). Dekel, A. and Aarseth, S. *Astrophys. J.* **283** (in the press).

11. Faber, S. M. *Proc. of the First ESO-CERN Symposium, Large Scale Structure of the Universe, Cosmology, and Fundamental Physics* (in the press).

12. Bond, J. R., Szalay, A. S. and White, S. D. M. *Nature* **301**, 584-585 (1983).

13. If the velocity dispersion data for the dwarf spheroidal galaxies are interpreted to imply heavy halos, the upper estimates in the figure result. The lower estimates follow from assuming that all the mass is visible. The former are probably more realistic, as discussed in Ref. 7, which is also the source of this figure and should be consulted for sources for the data displayed.

14. Aaronson, M. *Astrophys. J. Lett.* **266**, L11-L15 (1983). Aaronson, M. and Cook, K. *Bull. Am. Astron. Soc.* **15**, 907 (1983). Cook, K., Schechter, P. and Aaronson, M. *Bull. Am. Astron. Soc.* **15**, 907 (1983).

15. Faber, S. M. and Lin, D. N. C. *Astrophys. J. Lett.* **266**, L17-L20 (1983).

16. Lin, D. N. C. and Faber, S. M. *Astrophys. J. Lett.* **266**, L21-L25 (1983).

17. It is also possible, of course, that there are two or more forms of DM or that the DM is unstable or a decay product (possibilities discussed further in Roncadelli, M. and Primack, J. R. (SLAC-PUB-3304, 1984) and references therein).

18. Olive, K. A. and Turner, M. S. *Phys. Rev. D***25**, 213-216 (1982).

19. Pagels, H. R. and Primack, J. R. *Phys. Rev. Lett.* **48**, 223-226 (1982). Blumenthal, G. R., Pagels, H. and Primack, J. R. *Nature* **299**, 37-38 (1982).

20. Bond, J. R., Szalay, A. S. and Turner, M. S. *Phys. Rev. Lett.* **48**, 1636-1639 (1982). Bond, J. R. and Szalay, A. *Astrophys. J.* **274**, 443-468 (1984).

21. Preskill, J., Wise, M. and Wilczek, F. *Phys. Lett.* **120B**, 127-132 (1983). Abbott, L. and Sikivie, P. *Phys. Lett.* **120B**, 133-136 (1983). Dine, M. and Fischler, W. *Phys. Lett.* **120B**, 137-141 (1983).

22. Ipser, J. and Sikivie, P. *Phys. Rev. Lett.* **50**, 925-927 (1983). Sikivie, P. *Phys. Rev. Lett.* **51**, 1415-1417 (1983); *erratum* **52**, 695 (1984).

23. Ellis, J., Hagelin, J. S., Nanopoulos, D. V., Olive, K. and Srednicki, M. *Nucl. Phys. B*(in the press).

24. Carr, B. J. *Comments on Astrophys.* **7**, 161-173 (1978). Stecker, F. W. and Shafi, Q. *Phys. Rev. Lett.* **50**, 928-931 (1983). Freese, K., Price, R. and Schramm, D. N. *Astrophys. J.* **275**, 405-412 (1983).

25. Witten, E. (IAS preprint, 1984). Farhi, E. and Jaffe, R. L. (MIT preprint, 1984).

26. Vittorio, N. and Silk, J. (Berkeley preprint, 1984).

27. Bond, S. R. and Efstathiou, G. (preprint NSF-ITP-84-86, 1984).

28. Bardeen, J. (in prep., 1984).

29. Turner, M. S., Wilczek, F. and Zee, A. *Phys. Lett.* **125B**, 35-40 and (E) 519 (1983). Hara, T. *Prog. Theor. Phys.* **6**, 1556-1568 (1983).

30. Lynden-Bell, D. *Mon. Not. R. astr. Soc.* **136**, 101 (1967). Shu, F. H. *Astrophys. J.* **225**, 83 (1978).

31. Peebles, P. J. E. *The Large Scale Structure of the Universe (Princeton University Press, 1980). Efstathiou, G. and Silk, J. Fundamentals of Cosmic Phys.* **9**, 1-138 (1983).

32. White, S. D. M. and Rees, M. J. *Mon. Not. R. astr. Soc.* **183**, 341-358 (1978).

33. Rees, M. J. and Ostriker, J. P. *Mon. Not. R. astr. Soc.* **179**, 541-559 (1977).

34. See Ref. 7for a considerably more elaborate version of Fig. 4, with virialization curves for several multiples of the rms fluctuation spectrum for an open as well as an Einstein-deSitter universe, much more detailed galaxy and cluster data, and discussion of molecular and Compton cooling.

35. Geller, M. and Huchra, J. *Astrophys. J. Suppl.* **52**, 61-87 (1983).

36. Faber, S. M., Blumenthal, G. R. and Primack, J. R. *Astrophys. J. Lett.* (submitted).

37. Doroshkevich, A. G. *Astrophysics* **6**, 320-330 (1973).

38. The statistics of such correlations is discussed by Blumenthal, G. R., Faber, S. M. and Primack, J. R. (in prep., 1984).

39. Dressler, A. *Astrophys. J.* **236**, 351-365 (1980). Postman, M. and Geller, M. J. (CfA preprint 1939, 1983).

40. This point is mentioned in Ref. 7. The implications of inefficient galaxy formation are also discussed by Alex Szalay in these Proceedings.

41. Faber, S. M. *Proc. of the First ESO-CERN Symposium, Large Scale Structure of the Universe, Cosmology, and Fundamental Physics (in the press).*

42. *Efstathiou, G. and Jones, B. J. T. Mon. Not. R. astr. Soc.* **186**, 133 (1979). Fall, S. M. and Efstathiou, G. *Mon. Not. R. astr. Soc.* **193**, 189 (1980).

43. Sandage, A., Freeman, K. C. and Stokes, N. R. *Astrophys. J.* **160**, 831-844 (1970). Efstathiou, G. and Barnes, J. *Formation and Evolution of Galaxies*

and Large Scale Structures in the Universe (eds. Audouze, J. and Tran Th anh Van, J.) 361-377 (Reidel, Dordrecht, 1984).

44. Gott, J. R. and Thuan, T. X. *Astrophys. J.* **204**, 649-667 (1976).
45. Forman, W. and Jones, C. *Ann. Rev. Astron. Astrophys.* **20**, 547-585 (1982).
46. Geller, M. J. and Beers, T. C. *Proc. Astron. Soc. Pacific* **94**, 421-439 (1982).
47. Thuan, T. X. and Romanishin, W. *Astrophys. J.* **248**, 439-459 (1981). Cavaliere, A., Santangelo, P., Tarquini, G. and Vittorio, N. *Clusters and Groups of Galaxies* (eds. Mardirossian, F., Giuricin, G., and Mezzetti, M., R eidel, Dordrecht, Holland, in the press).
48. Davis, M., Efstathiou, G., Frenk, C. and White, S. D. M. (in prep., 1984).
49. Uson, J. M. and Wilkinson, D. T. *Astrophys. J. Lett.* **277**, L1-L4 (1984). Uson, J. M. amd Wilkinson, D. T. *Inner Space/Outer Space* (eds. Kolb, E. W. et al., Univ. of Chicago Press, in the press).
50. Rees, M. J. *The Very Early Universe*, (eds. Gibbons, G., Hawking, S. and Siklos, S.), 29-58 (Cambridge University Press, 1983).
51. This is discussed in more detail in Ref. 7and in the contributions of Alex Szalay and Avishai Dekel in these Proceedings.
52. Especially Refs. 2, 5, 6, and 7.

PANCAKES OF GALAXIES AND THE NATURE OF DARK MATTER

AVISHAI DEKEL

Institute for Theoretical Physics
University of California
Santa Barbara, California 93106
and
Department of Astronomy
Yale University
New Haven, Connecticut 06511

ABSTRACT

The distribution of galaxies on large scales is used to constrain the linear density perturbations at the onset of the gravitational instability era, and hence the nature of the dominant mass component and its density. The major theoretical scenarios, represented by N-body models, are confronted with the observations through a series of tests including the flattening of individual superclusters, the galaxy correlation function, the alignment of clusters, the null alignment of galaxies, and the cluster correlation function. Evidence is found for $\sim 30h^{-1}$Mpc "pancakes", elongated or flattened, which have formed only recently (at a redshift $z \sim 1$), slightly before rich clusters. Galaxies must have formed independently, earlier on. The linear perturbation spectrum is required to have a component with a coherence length at the pancake scale, and another component that has survived damping on galactic scales. There is some indication for $\sim 150h^{-1}$Mpc "superpancakes" of clusters which can be explained by yet another feature in the spectrum at that scale.

Einsten-deSitter cosmological models ($\Omega = 1$, $\Lambda = 0$) dominated by massive neutrinos or by "cold" particles face severe difficulties; the former cannot form galaxies in time and the latter fails to produce large scale pancakes and voids. A baryonic open universe, with $\Omega \sim 0.1$, can explain the required features in the perturbations spectrum and it provides a more promising picture. A comparable contribution ($\Omega \sim 0.1$) from "cold" particles may help forming galaxies and pancakes, and a similar contribution from neutrinos may help creating superpancakes.

1. Introduction

The observational data on the large-scale distribution of galaxies is now in a stage where certain general features can be isolated with some confidence and a better theoretical understanding of their genesis can be achieved via quantitative comparisons with theoretical scenarios. The relevant observations consist of extensive galaxy surveys on the two-dimensional sky, and of redshift surveys that provide estimated line-of-sight distances for limited samples which either cover large areas to a bright magnitude or penetrate deep in narrow fields. In our local neighborhood detailed information is provided by independently-measured distances. The scales of interest are in the range $1 - 100h^{-1}$Mpc (length scales are normalized to the present, and h is the Hubble constant in units of $100 km\, s^{-1} \text{Mpc}^{-1}$) corresponding to $1 - 10^4$ bright galaxies, *i.e.*, $10^{12} - 10^{16} M_\odot$. The dominant systems of galaxies are compact rich *clusters*[1] of $\sim 10^2$ bright galaxies in spheroids of a few Mpc, and *superclusters*[2] which are either larger associations of galaxies extending to a few tens of Mpc, like the Local Supercluster[3)–5)] or are groups of a few rich clusters extending to $\sim 100h^{-1}$Mpc[6)–9)]. The superclusters seem to be flat or elongated, with axial ratios up to 1:10, and therefore they are sometimes referred to as "pancakes." There is marginal evidence for a global cell structure[10),11)] where large volumes of low density ("voids", "holes") are surrounded by walls ("sheets", "pancakes") that intersect in denser linear structures ("strings", "filaments', "cigars"), which further intersect in knots — the cites of the richest clusters. The distribution of the rich clusters themselves provides important information on the structure on the very large scales of $\sim 100h^{-1}$Mpc.

It is assumed that the evolution of structure on the scales of interest is governed by gravity, and that the dynamical time scale is on the order of the Hubble time so that the presently observed structures are dynamically young enough to still memorize the initial conditions. I do not consider here cosmological scenarios in which nongravitational processes are involved in the formation of the large scale structure.

The theoretical scenarios start from early universe models which provide primordial density perturbations and certain families of particles that dominate the mass today. The perturbations evolve according to the nature of these particles,[13)] and the spectrum of perturbations can be predicted at the time when the universe turns matter dominated ($z \simeq 2.5 \cdot 10^4 \Omega h^2$) or when baryons and photons decouple ($z \simeq 10^3$), after which the perturbations can grow due to gravitational instability. Starting from a given spectrum at that epoch, the linear growth of structure can be followed analytically,[14),15)] and then the development of nonlinear structure requires numerical simulations using N-body techniques,[16),17)] which are accompanied when possible by crude analytical approximations to help understanding the dynamical processes.[18),19)]

My main concern here is the *quantitative confrontation of the theoretical scenarios with the observations*, which turns out to be a nontrivial task as visual comparison may be subjective and misleading. I will focus on *statistical tests* that are designed to distinguish between the major scenarios when comparing the simulated models with each other, and with the observed samples. The goal is to resolve the time sequence of formation, bottom-up versus top-down. The question is, in simple terms, which form first: galaxies, clusters, or superclusters? The answer will put constraints on certain features of the perturbations spectrum at $z \sim 10^3$, which in turn will constrain the nature of the dark matter and the universal density.

In § 2, I describe the theoretical scenarios to be considered, focusing on the origin of pancakes, and the numerical simulations. Subsequently, I describe and draw conclusions from the following tests: § 3. the galaxy-galaxy correlation function, § 4. alignments of clusters with superclusters, § 5. alignments of galaxies with pancakes, § 6. cluster-cluster correlations. Then, in § 7, I summarize and discuss my conclusions.

2. Theoretical Scenarios

a. Linear Spectra and Candidates for Dark Matter

As a working hypothesis, I consider three prototypes of linear density perturbation spectra at the beginning of the gravitational instability era: type I is *scale-free* and smooth on the relevant scales, type A has a critical *coherence length* λ, below which the perturbations have been damped out, and type AI is a *hybrid* of the two. The coherence length in A and AI is marked by a jump in amplitude over a small range of wavenumbers. On assuming scale-free primordial perturbations, many of the particles suggested as candidates for the dominant component of the mass in the universe admit spectra of perturbations at $z \sim 10^2$ that can be classified according to the above scheme in the following way:

(1) Baryons

(a) *Isothermal* perturbations, in which the baryons are perturbed on top of a uniform background of photons, do not evolve much prior to decoupling and result in a type I spectrum. If $\Omega h^2 > 0.03$, a feature is produced at the photon-baryon Jeans length before decoupling[15),20),21)]

$$\lambda_J \simeq (25 - 50)(\Omega h^2)^{-1}\text{Mpc}, \tag{1}$$

which is at the far end of the scales of interest here unless the baryons close the universe.

(b) *Adiabatic* perturbations, in which the photons and the baryons participate together, lead to a type A spectrum; photon diffusion and viscosity damp the acoustic oscillations below the critical Silk scale,[22)–29)]

$$\lambda_S \simeq 3(\Omega h^2)^{-3/4}\text{Mpc}. \tag{2}$$

(2) "Hot", weakly interacting particles, such as 10-100 eV *neutrinos*. A spectrum of type A is obtained because of free streaming of the particles when relativistic inside the horizon, below[30)–32)]

$$\lambda_\nu \simeq 14(\Omega h^2)^{-1}\text{Mpc}. \tag{3}$$

(3) "*Cold*", very weakly interacting particles, like axions, or particles more massive than 1 keV (*e.g.*, photinos, or right-handed neutrinos). The spectrum produced is like type I; it is very flat on galactic scales and below, and it steepens gradually on larger scales.[13),33)] In the case of "warm" particles, *e.g.*, having masses on the order of 1 keV, free streaming produces a relatively smooth damping length[32)] at $\lambda \simeq 1(\Omega h^2)^{-1}$Mpc, which is below the scales of interest here unless $\Omega \ll 1$.

A type AI spectrum could arise, in principle, from combinations of the above; adiabatic and isothermal perturbations in a baryonic universe, baryons and cold particles, neutrinos and cold particles, *etc.*.

The time sequence of formation is determined by the rms perturbation on a mass scale M, $\delta M/M$. On assuming for each component of the linear spectrum at $z \sim 10^2$, say, a power law

$$\langle|\delta_k|\rangle^2 \propto k^n, \tag{4}$$

one has

$$\delta M/M \propto M^{-\alpha}, \quad \alpha = (3+n)/6. \tag{5}$$

The slope, $-\alpha$, is more likely to be negative, so the evolution is expected to be ***bottom-up*** within each component. In particular, a spectrum of type I leads to hierarchical clustering, where the first objects to collapse are subgalactic, or galaxies, which later cluster non-dissipatively to groups, clusters, or perhaps superclusters.[34] If $|\alpha| \ll 1$, as is the case for cold particles on galactic scales, the structure over a wide range of scales form during a short cosmological time interval.[13),32),33),35] A spectrum of type A, on the other hand, leads to a *top-down* evolution, where structures just above the coherence scale and larger collapse first to superclusters, and smaller structures like clusters and galaxies form later due to nonlinear coupling of the large-scale perturbations and fragmentation processes. The hybrid, AI, starts evolving *bottom-up* on galactic scales, then superclusters collapse and induce the formation of rich clusters *top-down*, via nonlinear coupling.

b. The Origin of Pancakes

The presence of a coherence length (A or AI) leads to the formation of *pancakes*. This can be illustrated using the Zeldovich formalism in a Friedman universe. The Eulerian position of each particle, $\vec{r}$, as a function of its comoving position, $\vec{q}$, at a time t, is approximated by

$$\vec{r} = a(t)[\vec{q} - b(t)\vec{\psi}(\vec{q})], \tag{6}$$

where $a(t)$ is the universal expansion factor, and $b(t)$ is the universal linear growth rate; $b \propto a \propto t^{2/3}$ when $\Omega = 1$, and $b \to$ const. when $\Omega(t) \to 0$. The spatial dependence of the displacements is described by $\vec{\psi}(\vec{q})$, which is assumed to be derived from a potential, $\vec{\psi} = \nabla_{\vec{q}}\phi$, *i.e.*, no rotation. In the case of adiabatic perturbations the volocites are simply given by the time derivative of $\vec{r}$,

$$\dot{\vec{r}} = (\dot{a}/a)\vec{r} - a\dot{b}\vec{\psi}. \tag{7}$$

The density in the vicinity of each particle is obtained by differentiation with respect to $\vec{q}$,

$$\begin{aligned}\rho &= \bar{\rho}/\det(\partial\vec{r}/\partial\vec{q}) \\ &= \bar{\rho}/\det(\delta_{ij} - b\partial\psi_i/\partial q_j)\end{aligned} \tag{8}$$

where $\bar{\rho}(t)$ is the mean density in the universe.

If the perturbation is of the potential type, the deformation tensor $\partial\psi_i/\partial q_j$ is symmetrical and can be diagonalized locally giving principal axes $\hat{q}_j$, $(j = 1, 3)$. In the diagonalizing coordinates

$$\rho = \bar{\rho}(1 - b\lambda_1)^{-1}(1 - b\lambda_2)^{-1}(1 - b\lambda_3)^{-1}, \tag{9}$$

where $\lambda_j(\vec{q}) = \partial\psi_j/\partial q_j$ are the eigenvalues. Assume $\lambda_1 \le \lambda_2 \le \lambda_3$. If λ_3 is positive and has a local maximum $\lambda_{3,max}$ at the vicinity of some particle, the local density there increases as $b(t)$ grows and reaches a very high density at the finite time, t_c, defined by $1 - b(t_c)\lambda_{3,max} = 0$; the trajectories of neighboring particles along the q_3 axis intersect. If the perturbations are coherent over a length λ, the collapse would be systematic over a region of that size, and a planar "pancake" would form normal to the r_3 axis. It is highly probably that λ_3 is indeed larger than λ_2 and λ_1 such that the collapse will be essentially one-dimensional; Doroshkevich[36] estimated for a Guassian noise $\sim 40\%$ chance for $\lambda_3 > 0$ with $\lambda_1 \lambda_2 < 0$, but even for $\lambda_3 > \lambda_2$ a dynamical instability[37] would enhance the flattening. The pancake formation is especially efficient if the initial peculiar velocities are correlated with the position displacements, as in adiabatic perturbations. The initial anisotropies would determine the shape provided that pressure is negligible until late in the collapse.[38] This is indeed the case when $\sim 10^{15} M_\odot$ objects collapse while the Jeans mass is smaller by many orders of magnitude.

A formation of a cell structure naturally follows. In the intersections of two pancakes $\lambda_3 \sim \lambda_2 > 0$ and two-dimensional flows form dense linear strings. In the intersections of strings all three eigenvalues are positive and comparable to each other so that three-dimensional flows form rich clusters.

In the absence of a coherence length (I), on the other hand, local anisotropies are independently present on all scales, and any flattening would be a statistical accident and therefore not as pronounced or as persistent. Isothermal perturbations, which may have uncorrelated displacements and velocities, would tend to suppress any accidental coherency, but even adiabatic perturbations would not help much in this case. Furthermore, if an I spectrum starts clustering bottom-up from the post recombination Jeans scale, pressure would play an important role in resisting flattening early in the collapse of each system on all scales.

Thus, we associate the spectrum of types A and AI with *pancake scenarios*, and the spectrum of type I with *nonpancake scenarios*. A possible exception may be a very flat I spectrum, with a very significant power on large scales. It may behave effectively like an AI spectrum, giving rise to some pancaking. The "cold" particles spectrum is such an exception on scales below a few Mpc, where the lgoarithmic slope of the spectrum is $n \simeq -2$.

c. *Dissipative versus Nondissipative Pancakes*

The two types of pancake scenarios evolve very differently; the A spectrum leads to a *dissipative pancake* scenario[14],[39] while the AI spectrum gives rise to a *nondissipative pancake* scenario.[18],[40] In the A case, the baryonic material that collapses to pancakes is still gaseous because of the absence of perturbations below the coherence length. Gas dissipation keeps the pancakes thin while they fragment into galaxies. The gas collapsing to a singularity produces planar shocks that slowly propagate outwards. The later infalling gas is heated to $\sim 10^6$K, and is compressed behind the shocks, where radiative cooling to $\sim 10^4$K then allow fragmentation into subgalactic objects. Here *galaxy formation is triggered by pancake formation and occurs only there.* This puts a strict lower limit on the age of pancakes if one assumes that galaxies have formed before $z \sim 3$, as indicated, for example, by the existence of high redshift quasars, by the ages of globular clusters, by evolutionary models of galaxies, and by the need to hide newly formed galaxies. Also, all galaxies are expected to be found in pancakes or near them.

In the *nondissipative pancake* scenario[18] (AI) galaxies form first, as a result of the I component, and pancakes collapse later, when the A component above the coherence length becomes nonlinear. The epoch of galaxy formation is, therefore, independent of that of pancaking, and galaxies can be found away from pancakes as well. In this scenario, the collapse to pancakes is that of a gravitating system with no dissipation. A crucial point for the relevance of this scenario to the observed pancakes is that even with no dissipation, the pancakes remain very flat for a long time, as long as they continue to expand along the directions transverse to the first collapse. This can be illustrated by an idealized post-collapse model in which the bound particles oscillate about the pancake plane. Consider a particle trajectory $Z(t)$ normal to a uniform disk of radius $R(t)$. When the oscillation period, T, is shorter than the time scale associated with the expansion $R(t)$, there is an adiabatic invariant

$$\int_0^T \dot{Z}^2(t)\,dt \sim \bar{V}^2(t)T \sim \bar{Z}(t)\,\bar{V}(t), \tag{10}$$

where $\bar{Z}$ and $\bar{V}$ are some mean values of $Z(t)$ and $\dot{Z}(t)$, respectively, over one oscillation at $\sim t$. Since $\bar{V} \sim \bar{\mu}T$, where $\bar{\mu}$ is some mean value of the gravitational field exerted by the disk, $\mu(Z)$, the thickness is expected to vary like $\bar{Z} \propto \bar{\mu}^{-1/3}$. Now, the field near the disk, $Z \ll R$, is proportional to the surface density interior to $|Z|$, *i.e.*, $\bar{\mu} \propto R^{-2}$, and hence

$$\bar{Z} \propto \bar{V}^{-1} \propto R^{2/3}. \tag{11}$$

Thus, although the thickness of a pancake grows when it expands, it becomes relatively flatter in time as

$$\bar{Z}/R \propto R^{-1/3}. \tag{12}$$

This nondissipative cooling was confirmed by N-body simulations.[14)] It was found that a nondissipative pancake can be as flat and as cool as the observed local supercluster right after the first pancaking, and that it can actually become flatter later on, until subclusters grow inside the pancake and eventually determine its thickness and velocity dispersion.

d. Numerical Simulations

Most of the simulations used here are based on a comoving version of an N-body code[16)] which integrates directly the Newtonian equations of motion of the particles, with a chosen softening of the potential on small scales; the potential between the masses m_i and m_j separated by r_{ij} is assumed to be

$$\phi_{ij} = -Gm_i m_j/(r_{ij}^2 + \epsilon^2)^{1/2}, \tag{13}$$

where ϵ is the softening radius. This technique has the advantage of small scale resolution which is important for detailed studies of clusters. More bodies can be simulated on the expanse of resolution using Particle-Mesh codes, in which Fast Fourier Transforms are used to integrate the Poisson equation on a grid.[17)]

For the I scenario I use below a 4000-body simulation of Aarseth, Gott, and Turner,[41)] starting with a white-noise spectrum ($n = 0$). For the pancake scenarios, I use two 10^4-body simulations from a series of simulations by Dekel and Aarseth.[19)] These simulations started with an adiabatic spectrum that is a white noise on large scales and is sharply truncated below a critical damping length scale, λ. Model AI has, in addition, a component of white noise (isothermal fluctuations?) that dominates on small scales, with an A/I amplitude ratio ~ 3 at λ. In practice, the particles were first distributed uniformly inside a unit sphere, at the points of a cubic grid (model A) or, alternatively, at random (model AI). Then, the position and the velocity of each particle were perturbed by a superposition of $N_k = 1000$, small-amplitude plane waves,

$$\vec{\psi}(\vec{q}) = (2/N_k)^{1/2} \sum_{i=1}^{N_k} \sin(\vec{k}_i \cdot \vec{q} + \phi_i)\vec{k}_i/|\vec{k}_i|^2, \tag{14}$$

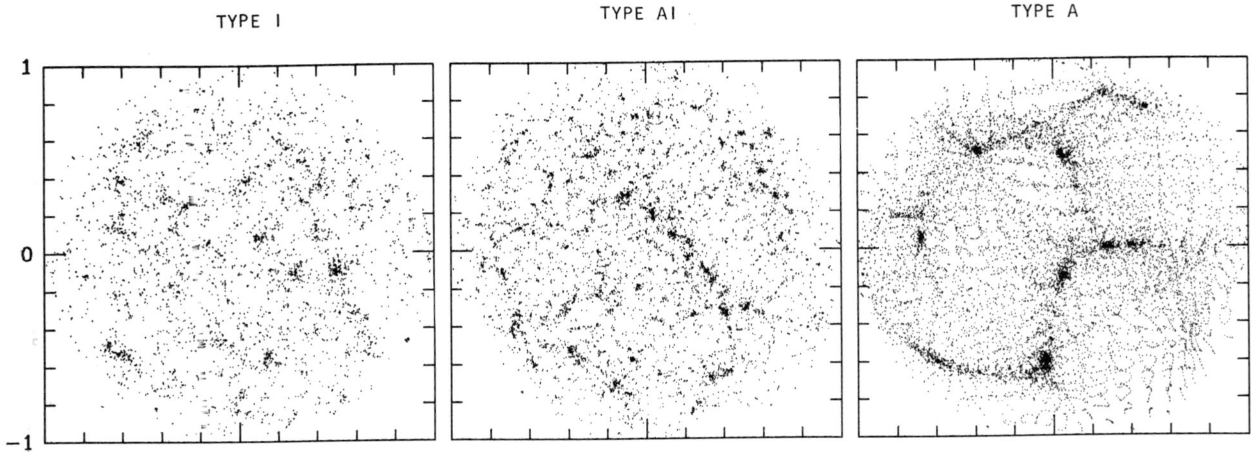

Fig. 1: Projections of the distribution of galaxies in the three basic models, I, AI, and A, with $N \simeq 4000$, 8000, and 10500 respectively. The clustering length ($r_o \simeq 5h^{-1}$Mpc) corresponds here to $\simeq 0.1$. The coherence length is $\lambda = 0.4$ and 0.6 in models AI and A respectively.

assuming random phases ϕ_i and wave numbers, $|k_i|$, that were chosen at random in the phase-space shell $k^3_{\min} < |k_i|^3 < k^3_{\max}$, where $k_{\min} = 2\pi$, and $k_{\max}$ corresponds to $\lambda = 0.6$ and 0.4 in A and AI respectively. All models assume an Einstein-de Sitter universe ($\Omega = 1$). The evolution of each system has been followed in the very linear regime by the approximation of Zeldovich [Eqs. (6) and (7)], until a stage where the rms density contrast was $b \simeq 0.25$. Then, the cosmological expansion factor was set to $a = 1$ and the N-body simulation started. The first pancakes reached singularities at a time stage corresponding to $a \simeq 4$.

The distribution of galaxies in the models at the times that correspond to the present epoch by a best fit of the correlation function are shown in Fig. 1 as viewed from one direction. A cell-like structure of filaments on the scale of λ can be recognized by the eye in A and AI and is absent in I. It is less pronounced in AI because, first, each line of sight crosses approximately five cells of superclusters and voids; second, the resolution is lower inside each cluster; and third, and most important, they are broken into rich clusters. The visual appearance of cold particles simulations[42] is more similar to that of the AI scenario than to the others; it shows some small scale filaments broken into clusters. The differences between the pictures are not always very pronounced, and they need to be quantified.

My approach is to look for tests that would clearly distinguish between the prototypical models, and then to use these tests to confront the models with the observations. The tests should be based on simple algorithms, and must be applicable observationally, or related to existing measurements.

3. The Galaxy-Galaxy Correlation Function

The two-point correlation function is by far the most common statistic used to measure clustering[15], and it provides a primary test for every model. Although it does not measure directly the shapes of structures, its evolution is different in the different scenarios, and there is a hope that some distinguishing features still show up in the present universe, before reaching the asymptotic shape which is common for all models later on, when rich clusters dominate the galaxy distribution.

The spatial correlation function is practically defined for a sample of N galaxies in a given volume by

$$1 + \xi(r) = N_p(r)/N_p^P(r), \tag{15}$$

where $N_p(r)$ is the number of pairs with a separation in the interval $(r-\Delta, r+\Delta)$, and $N_p^P(r)$ is the corresponding number of pairs expected in a Poissonian distribution of N particles in the same volume. More observable is the angular function, $\omega(\theta)$, which measures in analogy pair correlations on the two-dimensional sky. The latter is observed in extended samples to have a general power-law shape, which can be deprojected to give

$$\xi(r) \simeq (r/r_0)^{-\gamma}, \qquad \gamma \simeq 1.8, \qquad r_0 \simeq 5\mathrm{h}^{-1}\mathrm{Mpc}, \tag{16}$$

and a drop at $r \simeq 10\mathrm{h}^{-1}\mathrm{Mpc}$.

Redshift surveys allow a more direct evaluation of $\xi(r)$ by using the velocities as distance indicators. Figure 2 shows $\xi(r)$ as measured by Davis and Peebles[43] and by Einasto et al.[44] in the CfA catalog. It follows, in general, the power-law (16), but it shows, in more detail, a shoulder shape with an excess in the range $2 - 10\mathrm{h}^{-1}\mathrm{Mpc}$, where $1 + \xi(r) \propto r^{-0.85}$ to a

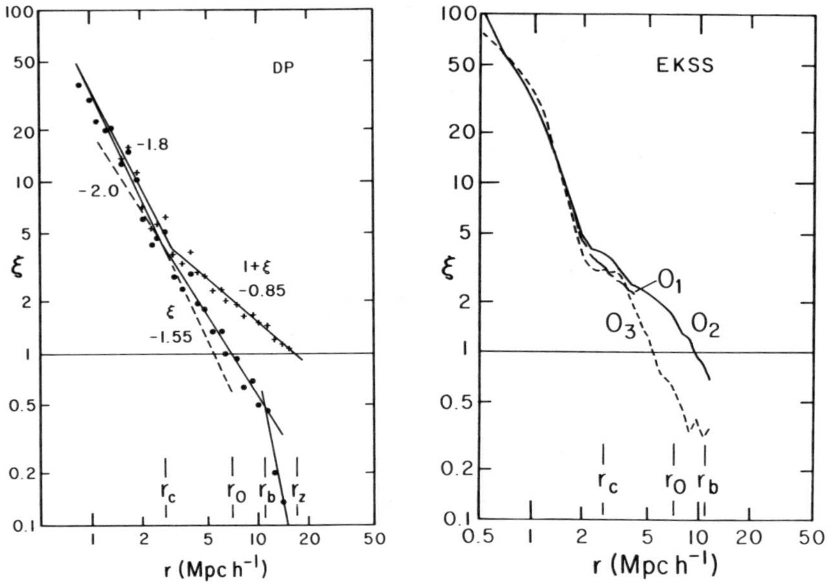

Fig. 2: The galaxy correlation function as calculated for subsamples of the CfA redshift survey. Left: The data for $\xi(r)$ (filled circles) is by Davis and Peebles[43] and the corresponding $1+\xi(r)$ (pluses) is added. The lines are best fits in the ranges $r < r_c$, $r_c < r < r_b$, and $r > r_b$ respectively, and their logarithmic slopes are marked. Right: $\xi(r)$ by Einasto et al.[44] for three volume-limited subsamples of which O_2 is the most reliable one.

good approximation. The lower limit of this range, r_c, coincides with the typical size of rich clusters, and the upper limit, r_b, with typical radii of pancakes. Dekel and Aarseth[19] argue that this excess is an imprint of the presence of pancakes, as I describe below.

The evolution of the correlation function in the I model is shown in Fig. 3a. It is self-similar in time, as expected from a scale-free spectrum,[45] and the expected slopes in the asymptotic regimes,[46] $\gamma_{\rm nonlinear} = (9+3n)/(5+n) = 1.8$ and $\gamma_{\rm linear} = (3+n) = 3$, are reproduced with $n = 0$. In the region of interest, $0.1 < \xi < 10$, the curve is convex with a steepening slope that clearly disagrees with the observed excess in $\xi(r)$.[43),47),48] Hence, an I scenario with $n = 0$ is ruled out. An I spectrum that has a negative power ($n = -1$, say) may better reproduce the observed shape of ξ near unity, but it may fail to reproduce its slope in the asymptotic nonlinear regime.

The effect of the presence of pancakes on $\xi(r)$ can be illustrated very simply as follows. Consider a cubic volume of size λ, with periodic boundary conditions, in which the matter has collapsed parallel to one side of the cube to a central thin plane of thickness d, where $d \ll \lambda$. Then in Eq. (15), for $d < r < \lambda$, $N_p(r) = N\rho\lambda\, 2\pi r\Delta$, while $N_p^P(r) = N\rho\, 4\pi r^2\Delta$, so that

$$1+\xi(r) = (r/r_z)^{-1}, \qquad r_z = \lambda/2. \tag{17}$$

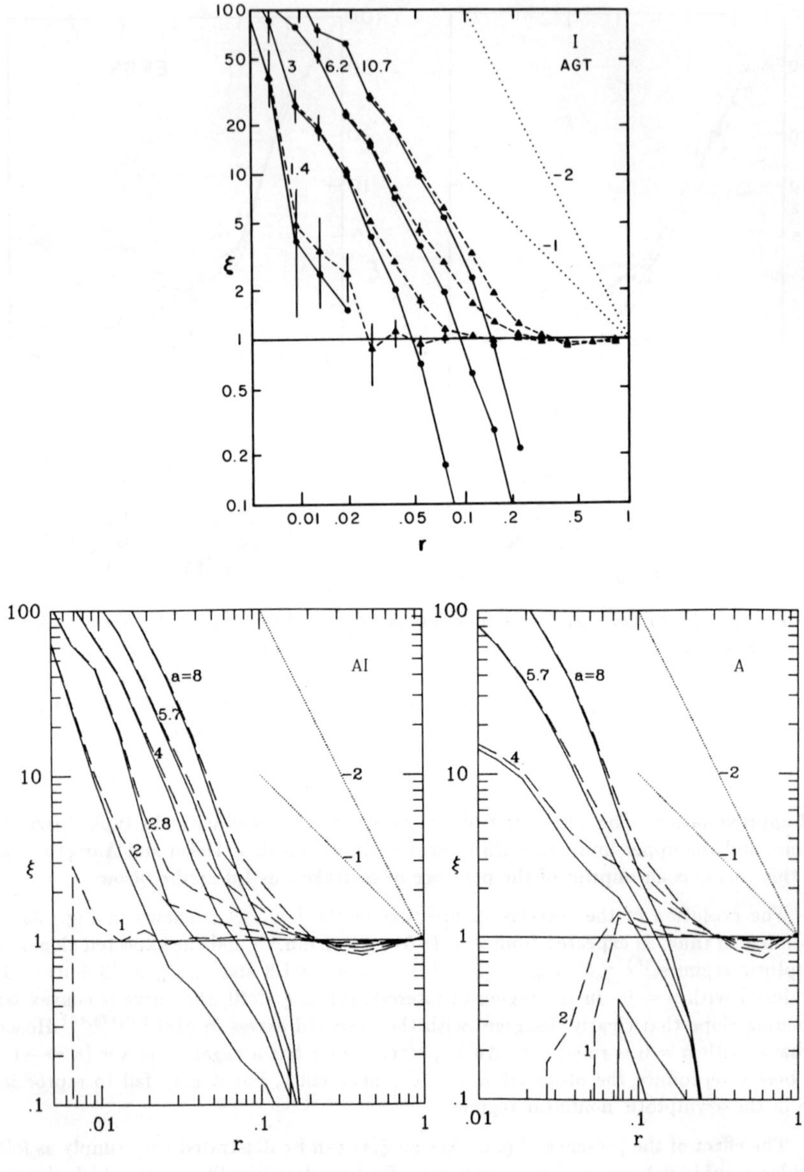

Fig. 3: The comoving evolution of $\xi(r)$ (solid) and $1+\xi(r)$ (dashed) in the N-body models I, AI, and A. Each pair of curves is characterized by the expansion factor. Pancaking occurs at $a \simeq 4$ ($b \simeq 1$). The coherence length is 0.4 and 0.6 in models AI and A respectively.

In the case of a two-dimensional collapse to a line one obtains in analogy

$$1 + \xi(r) = (r/r_z)^{-2}, \qquad r_z = \lambda/\sqrt{2\pi}. \tag{18}$$

Thus, $1 + \xi(r)$ has acquired a power law shape as a direct consequence of the transition from a three-dimensional system to a system of lower dimensionality, with no contribution from local gravity. The evolution of $\xi(r)$ is certainly not self-similar; dramatic growth and steepening occurs during the pancaking process, reaching an asymptotic shape that is fixed in comoving coordinates; for pure pancakes of size λ, one expects $\xi(r) = 0$ at $r_z \simeq \lambda/2$, and $\xi(r) = 1$ at $r_0 \simeq \lambda/4$.

The growth rate of $\xi(r)$ in a pure pancake model can be studied semi-analytically using the Zeldovich approximation with a perturbation of a single wavelength, λ, and in more detail using simple N-body simulations. These studies show a gradual growth of $1 + \xi(r)$ as a power-law which starts to deviate significantly from unity ($\propto r^{-0}$) at $a \sim \frac{1}{2}a_c$, and is very close to the asymptotic shape ($\propto r^{-1}$) already at $a \sim 2a_c$, where a_c is the expansion factor corresponding to the pancaking time t_c.

The effect of pancaking in a general A scenario can be studied first by using the Zeldovich approximation to follow kinematically the trajectories of the particles, starting from perturbations as described by Eq. (14), while the local gravity is turned off. The growth of $\xi(r)$ in such models, until $a \sim 2a_c$, was found to be strikingly similar to that in the pure pancake models, even though the spatial distribution of particles is now a complicated mixture of twisted, flattened objects. This similarity is because the members of most pairs with separations smaller than λ belong to the same "pancake" as a result of the large "voids" separating the pancakes. This indicates that the naive model of a pure pancake provides a useful approximation.

Figures 3b and 3c show the comoving evolution of $\xi(r)$ (solid) and $1 + \xi(r)$ (dashed) in models AI and A, where each pair of curves is characterized by the corresponding expansion factor (recall: pancaking at $a \simeq 4$, $b \simeq 1$). At each stage there is a characteristic radius, r_c, at which $1 + \xi(r)$ shows a concave break. In the range $r \geq r_c$ the growth in both models is as expected from pancaking; the function grows gradually to $1 + \xi(r) \propto r^{-1}$ while the zero of ξ is fixed at $r_z \simeq \lambda/2$. In the range $r \leq r_c$ in model AI, the slope is $\gamma \sim 2$ from early stages, and $\xi(r)$ evolves self-similarly as in model I. In model A, the initial growth of $\xi(r)$ on small scales is due to nonlinear coupling of the large scale perturbations at the lines and knots of intersection between neighboring pancakes. Once originated, the small-scale perturbations continue to grow subject to their own gravity, reaching a correlation function similar to that of models AI and I at $a \sim 2a_c$. In both models, when the small scale clustering develops, r_c propagates to larger comoving scales. Eventually, after $a \sim 2a_c$, $\xi(r)$ approaches a shape very similar to that in the I scenario, *i.e.*, too steep near unity; the subclusters grow to the pancake size and take over. From then on, the scenarios are indistinguishable, and are all incompatible with the observations. Thus, the observed correlation function and, in particular, its shoulder shape near unity, is reproduced by the pancake models at $a \sim 1.5a_c$ ($b \sim 1.5$), when $r_0/\lambda \simeq 0.2$. Similar results were obtained in other simulations of the A scenario,[49)–51)] some with a more accurate neutrino spectrum.

The conclusions from a detailed comparison of the observed galaxy correlation function with that of the models, including open universe models, can be summarized as follows:

(1) Hierarchical clustering (I) alone is not enough, unless, perhaps, the spectrum is very flat on large scales.

(2) There is a coherence length at

$$24 < \lambda < 34h^{-1}\text{Mpc}. \tag{19}$$

Silk damping with $\Omega_b \sim 0.1$, or neutrinos with $\Omega_\nu \sim 1$, can provide an appropriate feature.

(3) Pancaking must have occurred very recently, at

$$z_c \leq \begin{cases} 0.5 & \text{if } \Omega = 1 \\ 1.9 & \text{if } \Omega = 0.1 \end{cases} \tag{20}$$

Thus, *the pancakes are young dynamically.* This means that they cannot be responsible for galaxy formation, and it provides a strong argument against the A scenario, and in support of the AI scenario.

(4) An I component is needed to provide subclustering and galaxy formation in time, with I/A amplitude ratio of ~ 0.5.

If the galaxies do not follow the mass distribution in the pancake scenarios the results are only slightly changed. If galaxies preferentially form in dense regions, the slope of $\xi_g(r)$ becomes steeper and r_0/λ becomes bigger at any given time, so the age constraints are even tighter. The constraints are slightly loosened if galaxies form in flat pancakes, but not in denser strings and clusters.[52]

The "cold" particle spectrum provides a correlation function which grows rapidly to the required slope. However, the predicted correlation length, r_0, is on the order of only $\sim 1(\Omega h)^{-1}h^{-1}$Mpc at that time. This is consistent with the galaxy correlation length only if $\Omega h \sim 0.2$.[42] Alternatively, an $\Omega = 1$ model can be repaired if galaxies formed selectively only in regions of initial density above some threshold, which would boost $\xi_g(r)$ up according to the actual height of the threshold.[42],[53],[54] The prediction in this case is that $\xi_g(r)$ would not evolve much once galaxies have formed.

A change of slope at moderate redshifts is observable, in principle. There are marginal indications[55],[56] for a slight flattening of $\omega(\theta)$ for galaxies in the magnitude interval $23 \leq J \leq 24$ corresponding roughly to $0.5 \leq z \leq 0.7$. However, larger and deeper surveys are required in order to measure $\omega(\theta)$ reliably beyond $z \sim 0.5$, and obtain firm conclusions. The fact that the Lyman-alpha absorption clouds, along the lines of sight to quasars, show no measurable spatial correlations at $z > 1.7$[57)–59] may be of relevance. If they cluster like galaxies, it would indicate a rapid growth of ξ_g recently[57] in agreement with the AI scenario but not with the A scenario. It would be inconsistent with the I scenario as well,[54] where the evolution is at a constant, slow rate. If ξ_g indeed evolves at recent epochs, the "cold" particle scenario would survive only if $\Omega h \sim 0.2$; the selective galaxy formation would not work.

4. Tests for Pancakes; Alignment of Clusters and Superclusters

a. Alignment of Abell Clusters

Rich clusters of galaxies can be used to probe the elongation of structures on larger scales.[60] Most of the clusters show well defined position angles of their major axes,[61],[62] and at least in two major superclusters, Coma and Perseus, the cluster position angles are known

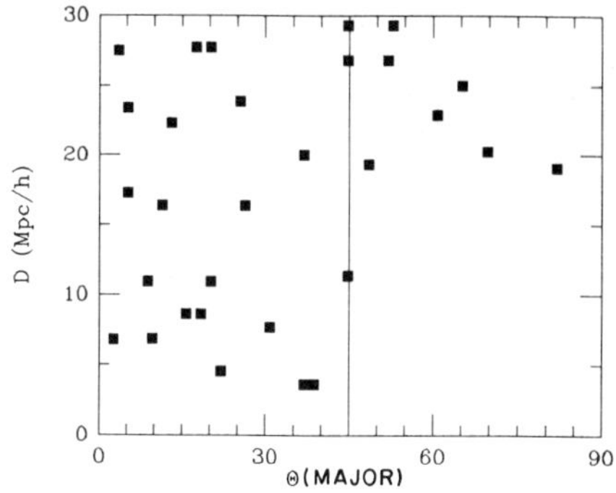

Fig. 4: Binggeli's effect.[63] The parameter θ is the angle between the major axis of an Abell cluster and the line connecting its center of mass to that of its nearest neighboring cluster, and D is the spatial distance between the clusters.

to be aligned with the supercluster major axes.[2] Binggeli[62] has given more significance to this effect by finding that clusters of galaxies in general "tend to point toward each other." For each of the 44 Abell clusters he studied, Binggeli has determined the position angle of the major axis, using the 50 brightest members of the cluster, and the position angles of the lines connecting its center of mass to the centers of mass of neighboring clusters. In Fig. 4 we show Binggeli's results for nearest neighboring clusters, where θ is the angle between the position angles described above, and D is the three-dimensional separation of the clusters, obtained using their redshifts. There is a striking alignment of the orientations on scales below $20h^{-1}$Mpc which extends at least out to $50h^{-1}$Mpc. When only the nearest neighbors are considered, $\langle\theta\rangle = 29.5^o \pm 4.4^o$. When all neighbors are considered, $\langle\theta\rangle = 36^o \pm 5^o$ and $40^o \pm 2^o$ for $D < 25h^{-1}$Mpc and $< 50h^{-1}$Mpc, respectively. All these means are significantly below 45°. Unlike many other cases in this field, there is no need here for any fancy statistics to see the effect, and it has been demonstrated by Binggeli, using Monte Carlo techniques, that the errors made in determining θ are indeed very small.

The possible interpretation of Binggeli's alignment is apparently ambiguous: while such an effect may be a natural consequence of the large-scale *flattening* in the pancake scenarios, where anisotropic shapes and velocity dispersions in the clusters were induced by the anisotropic collapse of the parent supercluster, it may alternatively be a result of *tidal* interactions between protoclusters which act in the I scenario as well. This assertion is based on the estimates[63],[64] that mutual tides in the protocluster stage are indeed capable of inducing prolate shapes that point to each other.

In order to identify the actual source of the alignment, we have applied an analysis similar to Binggeli's (1982) to the clusters we identify in the N-body models described above. The procedure adopted for identifying clusters in the simulations is as follows.[65] For a given value of a separation parameter d, every two particles that are separated by less than d are defined as "friends," and then all mutual friends are assigned to the same cluster. The parameter d determines the richness of the clusters identified and is closely related to the overdensity at the outer parts of the clusters. The values of d chosen to identify rich clusters in the present context correspond to an overdensity greater than ~ 35. A requirement of a minimum number of members (typically ~ 30) was imposed in order to focus on the

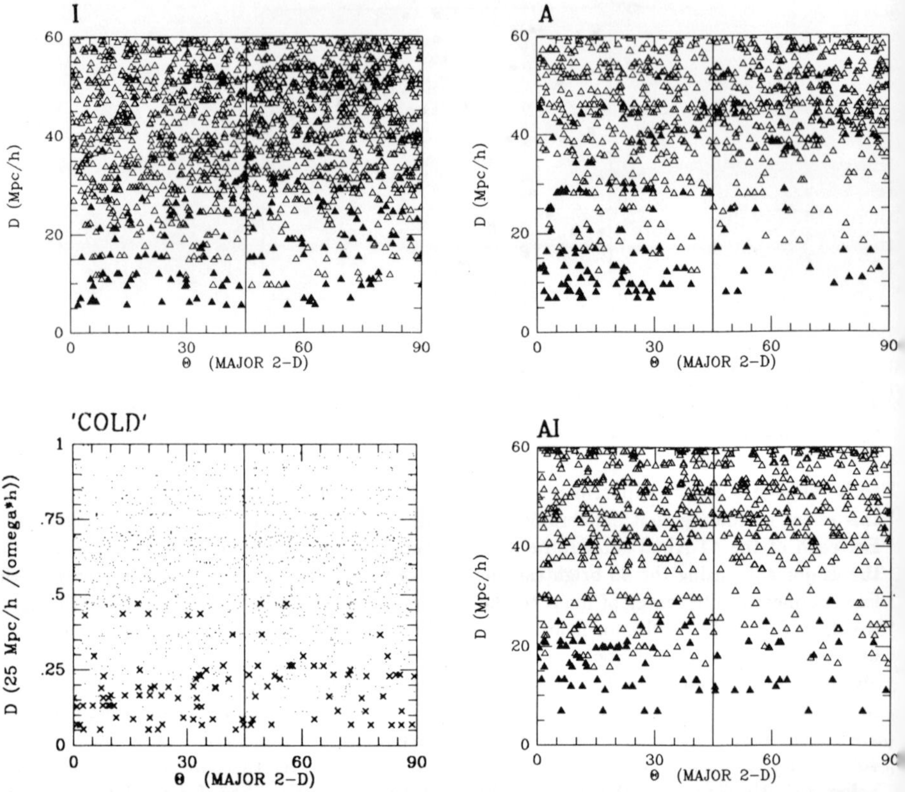

Fig. 5: Cluster-supercluster alignments in the simulations projected into two dimensions. θ and D are as in Fig. 4. Filled symbols (or x) correspond to nearest neighbors and the rest to any neighbors. The mean angles in the different models for all neighbors with $D < 30h^{-1}$Mpc are: $\langle\theta\rangle_I = 43^o.8 \pm 1.4^o$, $\langle\theta\rangle_A = 29^o.8 \pm 1.7^o$, $\langle\theta\rangle_{AI} = 35^o.7 \pm 2.1^o$.

richest clusters that would resemble the Abell clusters as much as possible. Following the above procedure, we find in each simulation typically a few tens of clusters which contain $\sim$ 10% of the total mass, $\sim 10^{15} M_\odot$ each. Once a cluster has been identified, its principal axes of inertia were computed from the distribution of galaxies within it, by solving the corresponding eigenvalue equation. The principal axes were computed in three dimensions and in three orthogonal two-dimensional projections. All clusters were flat enough for their principal axes to be determined uniquely, and their directions were found to be insensitive to the method used to identify the clusters. Then the angle θ (in projection and also for each principal axis in 3D) and the 3D distance D were determined for each cluster-neighbor pair.

Figure 5 shows the results in projection, to be compared to Binggeli's data. Each of the plots marked I, A, and AI is a superposition of the results from a few realizatons of each of the basic models, based on the simulations described in § 2 together with simulations run by Frenk, White, and Davis.[48] The time stage is determined by the shape of the correlation function, and the scaling of D is determined by equating the correlation length, r_0, to $5h^{-1}$Mpc at that time.

It is clear that there are no alignments in the I model. The analytical estimates[63],[64] overestimate the tidal effect because they assume for the protoclusters overdense perturbations above a *uniform* background. However, when realistic protoclusters start contracting relative to the expanding background, they are likely to evacuate underdense regions around them, which tend to weaken their gravitational influence on other protoclusters. It is evident that such underdense regions are present between the clusters in the I simulations.[66] This is, again, a strong argument against an I scenario (with $n \sim 0$).

On the other hand, the A model shows an alignment which is even more pronounced than the observed effect, and the AI model still shows a significant effect which is comparable to that observed. Thus, the observed alignment provides an evidence for a *pancake* scenario (A or AI), with a coherence length at $\lambda \simeq 30h^{-1}$Mpc, in agreement with Eq. (19).

The alignment could have come about only if superclustering was already in progress when the clusters formed. This is an argument for top-down evolution from superclusters to clusters, and against bottom-up scenarios in general. If superclusters are indeed young dynamically as concluded above (§ 2, § 3), then rich clusters must also be young; they should show dynamical evolution at $z \sim 1$. This is in agreement with rough constraints obtained from the present overdensity in Abell clusters ($\delta\rho/\rho \sim 200$ inside an Abell radius of $1.5h^{-1}$Mpc for richness $R = 1$) using a spherical model in an $\Omega \sim 1$ universe. The current null observation of evolution in clusters for $z < 1$ argues for $\Omega < 1$ [Eq. (20)], but evolution must be observed at some epoch $z < 2$ if our interpretation of the cluster alignment is correct.

In order to complete the set of models, we have analyzed two new "cold" particles simulations, with 32^3 particles, by Davis *et al.*.[42] The results are shown in Fig. 5 ('cold'), where D is given in units of the side of the comoving cubic volume of the simulation, which corresponds to $32.5(\Omega h^2)^{-1}$Mpc. The time stage is chosen when the slope of the mass correlation function is $\gamma \simeq 1.8$. An alignment comparable to the observed effect (for $D < 30h^{-1}$Mpc) is detected only for $D < 0.2 \cdot 32.5(\Omega h^2)^{-1}$Mpc. Thus, the pancaking in the "cold" particle model is on very small scales, and is compatible with the observatons only if the universe is open, with $\Omega h \leq 0.2$. A similar conclusion is obtained when comparing the mass correlation length in these simulations to the observed galaxy correlation length.[42] It is argued that selective galaxy formation in high density peaks[53],[54] can bring $\xi(r)$ in an $\Omega = 1$ model to an agreement with the observations, but the scale deduced from the alignment of clusters should not be affected by such selective galaxy formation. It, therefore, poses a severe *difficulty* for a "cold" particle scenario with $\Omega = 1$.

My main conclusions are then:

(1) The alignment is due to large scale pancaking. It supports pancake scenarios (A or AI) with a coherence length at $\lambda \sim 30h^{-1}$Mpc.

(2) Tidal effects in an I ($n = 0$) scenario do not produce a significant alignment.

(3) The "cold" particle spectrum may provide a comparable effect only if $\Omega h \leq 0.2$.

(4) Clusters should form after, or during, pancake formation.

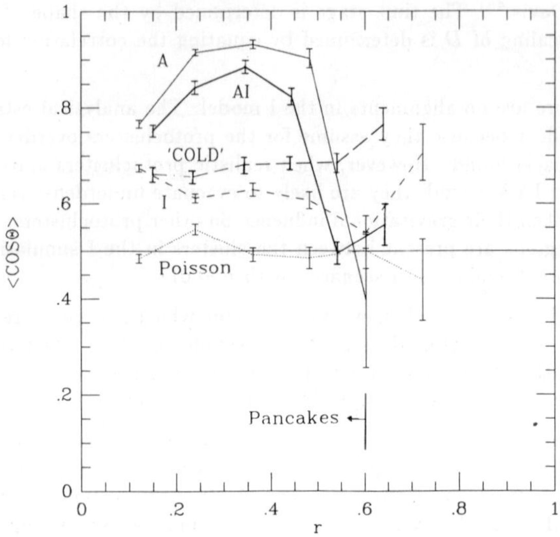

Fig. 6: Alignments on general scales in the models (preliminary results).

b. Alignments on General Scales

Motivated by the success and simplicity of the cluster alignment test in which one simply measures well defined position angles, we[67)] have generalied it to a *general statistical test* as follows: For any given length scale r, the test measures the alignment of the distribution of galaxies on scales smaller than r with that on scales larger than r. About each galaxy we center two spherical shells (or circular rings in 2D), one between the radii $r - r_2$ and $r - r_1$, and the other between $r + r_1$ and $r + r_2$ (typically $r_2 - r_1 = r_0$, the clustering length, and $r_1 = 0.5r_0$). We then find the principal axes of the distribution of galaxies in each shell and calculate the angles between the corresponding major, intermediate, and minor axes, θ_1, θ_2, and θ_3. We repeat the procedure about every body in our sample and calculate the mean of θ_i as our statistic. In Fig. 6 we show preliminary results for the major axes alignments in the basic models I, A, and AI, and in the "cold" particle model, in comparison with a Poissonian distribution. The models are scales to have the same correlation length, r_0. The distinction between the pancake models and the non-pancake models on scales below the coherence length is remarkable! This test seems to distinguish between the models better than any other that we have studied so far. This is not trivial because many tests that looked promising *a priori* failed to distinguish quantitatively between the models either because of a false signal in the non-pancake models, nontrivial sensitivity to irrelevant parameters, *etc.*(see below).

The proposed test is simple and cheap to calculate, which promises to make it useful in future studies by observers or theoreticians as a complementary statistic to the two-point correlation function. Its applicability in 2 or 3 dimensions, and on all clustering scales, is a very appealing feature, and the fact that it works so well in the limit of Abell clusters (§ 4a) is very promising. We are in the process of optimizing the test, investigating its statistical properties, and applying it to extensive 2D and 3D data samples.[68)]

c. *Other Tests for Pancakes*

Various statistical tests for pancakes, or a filamentary structure, have been proposed in recent years. Some[68)–70)] detected a certain signal in the galaxy counts of Shane and Wirtanen from the two-dimensional Lick survey, and others[71)–74)] looked for filaments in other data samples. These tests have not been calibrated by realistic simulations so it is difficult to assert from them quantitative conclusions concerning "pancakes" in the context of the competing theoretical scenarios. Three-dimensional tests based on axial ratios of individual structures,[44)] or on the multiplicity function of clusters, have indicated pancaking in the CfA redshift sample in the vicinity of the Local Supercluster, but only still on a qualitative level.[60)]

The Soviet group[38),44),75)] have suggested a test based on *percolation theory*; they found it to be sensitive to the differences between numerical models and concluded that the comparison with the CfA data favors the pancake scenario over the I scenario. However, we[76)] found severe difficulties in the use of this test by studying the detailed N-body models along with simple toy models. The percolation properties were found not to be very sensitive to the presence of "pancakes" and "strings" once they are *clumpy*, and hence they do not distinguish properly between the models I,[77)] AI, and "cold",[78)] even in comparison with a Poisson (P) distribution. In the case of very smooth pancakes (A), the ability to percolate depends on sampling parameters, such as the mean *number density* and the *volume*, in a way which is unknown a priori because it depends on the same properties that the test ought to measure. This problem, could, in principle, be eased by using limited samples of high mean number density (an order of magnitude denser than the CfA redshift survey volume limited at 4,000 km s^{-1}) and by comparing to models of identical number density and volume. We[76)] suggested an alternative approach, based on the sampling effects themselves that may provide a qualitative test for pancakes in samples of lower densities, and actually confirms that the local superclusters is a flat pancake. However, a great deal of caution is still called for when applying the percolaiton test, and the conclusion should be regarded as qualitative accordingly.

5. Alignment of Galaxies and Pancakes

If galaxies formed in pancakes according to the dissipative scenario (A), one expects some sort of correlation between the orientation of a galaxy and the orientation of its parent pancake. The fragmentation process in pancakes is not yet fully understood so it is hard to predict the exact form of this correlation, but some preliminary ideas have been discussed.[36)] For example, the average bound volume around a high density maximum in an initial Gaussian noise is an ellipsoid whose principal axes coincide with those of the deformation tensor. Thus, an elliptical galaxy, whose shape is determined by the initial anisotropy in position and velocity, is expected to be aligned with its parent pancake.

On the other hand, the angular momentum of a galaxy is expected to lie in the pancake plane. If the principal axes of the Lagrangian volume that would collapse to a galaxy coincide with those of the pancake, then the mean angular momentum vanishes, but its dispersion is zero only along axes of symmetry of the protogalaxy. If the principal axes do not coincide, then the average angular momentum is nonzero. A planar shock would then amplify the components of the angular momentum that lie in the pancake plane because of the reduction in the normal velocity behind the shock, and the associated change in position. Thus, the minor axis of a spiral galaxy, whose shape is determined by rotation, is expected to lie in the plane of its parent pancake, *i.e.*, their major axes are expected to be perpendicular.

The north pole of the Milky Way indeed lies in the plane of the local supercluster. However, more systematic searches for such alignments in various clusters and superclusters came up with null results. The brightest galaxies in rich clusters are often aligned with their parent clusters,[61),62)] but this is believed to be an exception which is probably due to evolution. Normal galaxies are typically not aligned with their parent clusters, although occasionally a marginal alignment is detected.[79)–81)] In particular, an alignment reported in the Coma cluster[82)] is probably a result of a systematic error. All the efforts to detect alignments in the local supercluster[83)–86)] came up with null results.

I[87)] recently pursued a search for galaxy-pancake alignment in two extensive galaxy catalogs that cover the north sky and most of the south sky: the Uppsala (UGC) and the ESO-Uppsala (EUC) catalogs. They are suitable for such a search because they list the position angle of each galaxy on the sky together with the common information on the sizes of its major and minor axes, its type, *etc.*. These catalogs are limited by galactic diameters to $> 1'$, which corresponds to a typical L^* galaxy at $\sim 40h^{-1}$Mpc. This is ideal for our purpose because it is deep enough to encompass a few superclusters, but shallow enough to avoid significant smearing due to superclusters that overlap in projection.

The statistic used is similar to that described in § 4. To test an alignment on a scale of separation on the sky S ($S = 5^o, 10^o, ..., 90^o$), I draw around each galaxy a ring of radii S and $S + \Delta S$ ($\Delta S = 5^o$), and calculate the major axis of inertia for the distribution of galaxies in that ring. The angle θ between this axis and the position angle of the galaxy at the center is calculated, and its average, $\bar{\theta}(S)$, over all galaxies as centers, is the required quantity. The analysis can be applied to the whole catalog, or to various subsamples. In particular, it is useful to select galaxies that belong to prominent elongated structures according to an overdensity criterion using a cluster finding algorithm (see § 4).

Before testing the observed catalogs, I have created an imitation of a dissipative pancake model using the A model (§ 2d). First, I tagged particles as galaxies only if they were in pancakes at $a = 4$ ($b = 1$) according to the Zeldovich approximation. In practice, I calculated for each particle the eigenvalues of the deformation tensor [Eq. (9); $\lambda_1 \leq \lambda_2 \leq \lambda_3$] and selected only particles that had $\lambda_3 \geq 1$. Their positions are taken from model A at $a = 5.7$. For each of the 864 galaxies selected, I then assumed that its minor axis lies in the pancake plane, along the longest principal axis that corresponds to λ_1. Finally, a 2D catalog was created by projecting the system onto the sky of an observer located at the center of the simulated sphere, and calculating the position angles of the galaxies assumig thin disks. The depth of this model catalog is comparable to the observed ones.

Figure 7 shows the galaxy-pancake alignment in this dissipative pancake model (left) by $\bar{\theta}(S)$. The error bars correspond to $\pm\sigma$. A significant anti-alignment is detected in the model below $S \sim 45^o$, which indeed corresponds to the coherence length λ at the given catalog mean depth. As a check for the algorithm, the same analysis was applied to galaxies at identical positions, whose position angles were assigned at random (right).

The analysis was applied to a variety of subsamples of the observed catalogs. I looked at all prominent structures, together or separately, at various levels of density contrast, at various depths, and for various galaxy types. Figure 8 shows the results from two of these

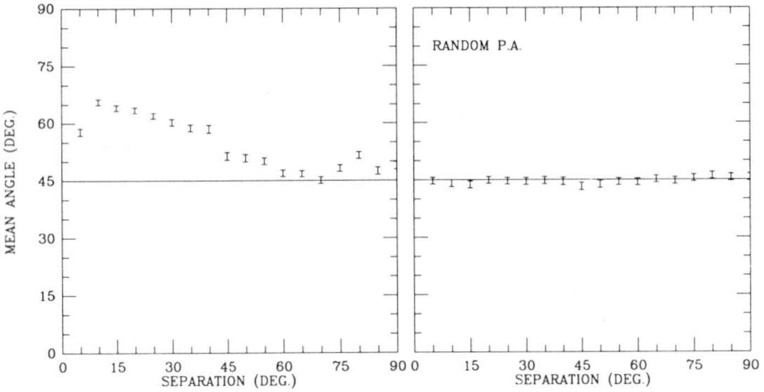

Fig. 7: Galaxy-pancake anti-alignment in a dissipative pancake model. The mean alignment angle, $\bar{\theta}$, is plotted as a function of sky separation, S. Right: same galaxy positions but with random position angles.

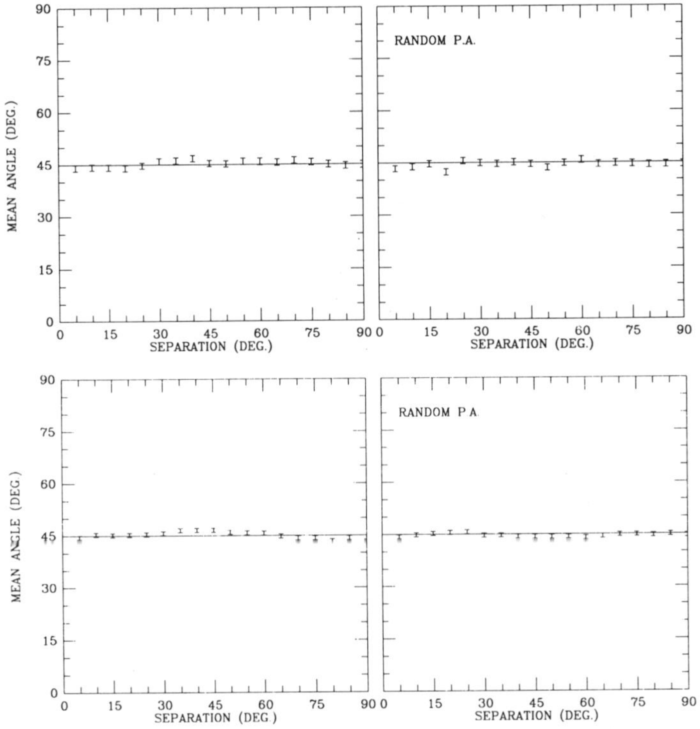

Fig. 8: Galaxy-pancake alignment in the subsamples shown in Fig. 9. Up: UGC, bottom: EUC. The notations are as in Fig. 7.

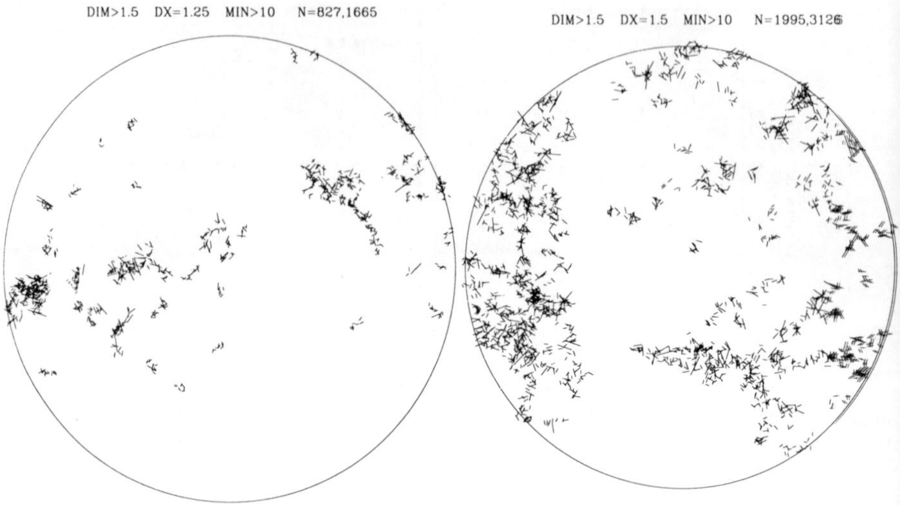

Fig. 9: Projections of clustered subsamples from UGC (left) and EUC (right). Right Ascension grows counterclockwise and Declination is respectively 0° and $-17^o30'$ at the circumference and $+90^o$ and -90° at the center.

subsamples, which are shown in projection in Fig. 9. Only galaxies with diameters $> 1.5'$ were included in these particular subsamples. The separation parameters used to identify clusters here were $d = 1.25^o$ and 1.5^o for UGC and EUC respectively. The position angles of only those galaxies with an axial ratio $< 2/3$ were considered in the analysis.

In all cases but one I found no significant evidence for alignment on any scale. This conclusion applies, in particular, to the Local Supercluster and to the Perseus-Pisces supercluster. In one, small, high-density subsample of the southern counterpart of the Local Supercluster, there was a marginal signal of anti-alignment on the 2σ level (note that the $\bar{\theta}$ values obtained for different separations s are not independent of each other). This signal weakens when a larger, less dense subsample of the same supercluster is tested.

I conclude that the data is consistent with *no alignment of galaxies and pancakes.* This argues against top-down scenarios in which galaxies form in pancakes (*e.g.*, *A*), and in favor of galaxy formation before clusters and superclusters.

6. Cluster-Cluster Correlations; Superpancakes

A very valuable information on very large scales is provided by the correlation function of rich clusters of galaxies, $\xi_c(r)$, which is observed to be in excess of the galaxy correlation function, $\xi_g(r)$; $\xi_c = 1$ at $r_c \simeq 25h^{-1}$Mpc for Abell ($R \geq 1$) clusters as analyzed by Bahcall and Soneira[88], and at $r_c \simeq 18h^{-1}$Mpc for less rich clusters as measured by Shectman,[89] while $\xi_g = 1$ at $r_g \simeq 5h^{-1}$Mpc [Eq. (13)]. In order to find out whether this effect is reproduced in the standard clustering scenarios, Barnes, Dekel, Efstathiou, and Frenk[90] have applied cluster finding algorithms to cosmological N-body simulations, comparing ξ_c for clusters of a

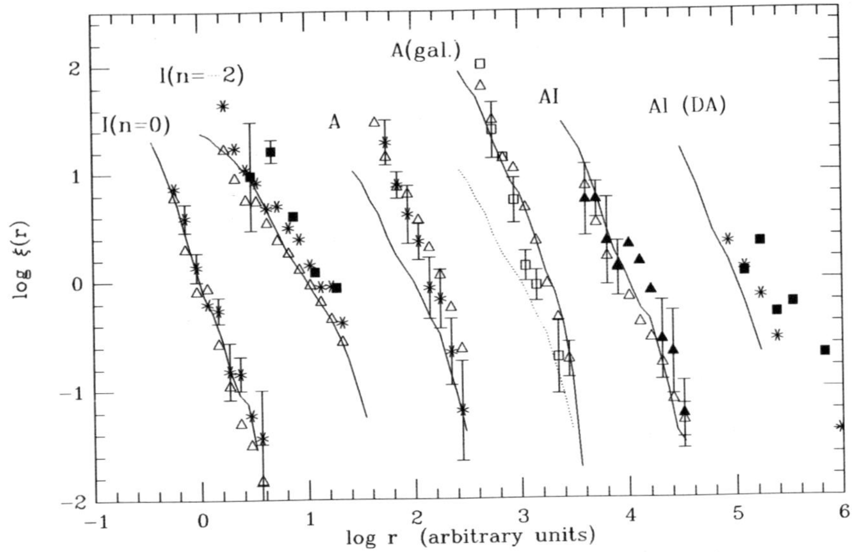

Fig. 10: $\xi_g(r)$ (solid curves) and $\xi_c(r)$ (symbols) in N-body models.[90] The symbols in order of growing richness are: open triangles, asterisks, filled squeres, and filled triangles.

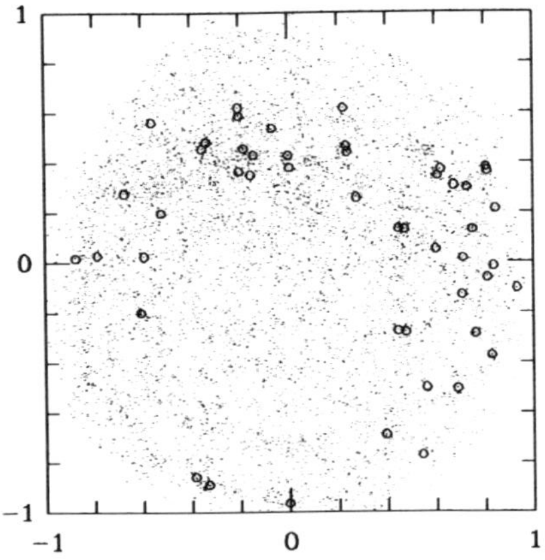

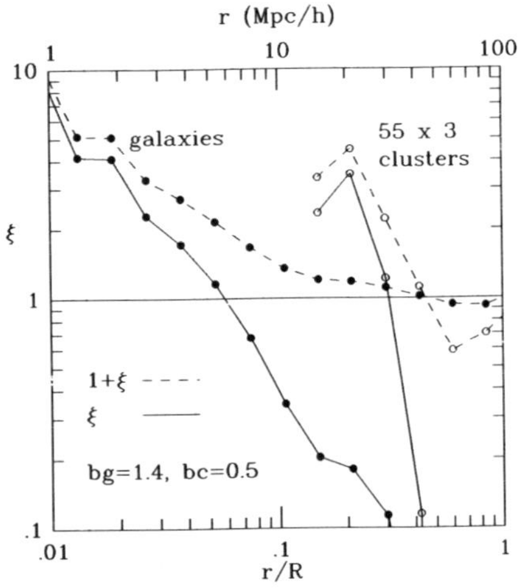

Fig. 11 (left): A projected distribution of galaxies (dots) and rich clusters (circles) in a model with two coherence lengths for pancakes ($\lambda_g = 0.2$) and superpancakes ($\lambda_c = 1$).[93]

Fig. 12 (right): $\xi_g(r)$ and $\xi_c(r)$ averaged over three models like the model shown in Fig. 11.

certain richness to ξ_g as given by the individual bodies when the slope of ξ_g is $\gamma \simeq 1.8$. Figure 10 summarizes the results. In I ($n \sim 0$) models, ξ_c and ξ_g practically coincide. In pancake scenarios (also Ref. 91), ξ_c is steeper than ξ_g, but $r_c/r_g \leq 2$, with no richness dependence. In the case of a very flat spectrum with $n = -2$, there is a significant growth of ξ_c with richness, but at a rate which seems weaker than the observed rate; while the richest clusters analyzed in the model are comparable to Shectman's, the excess detected in the model is smaller, and a naive extrapolation to Abell-like clusters indicates $r_c/r_g \sim 2$ only. At an earlier time, when ξ_g is too flat with $\gamma \simeq 1.4$, the excess in the model comes close to Shectman's, but the extrapolation to Abell clusters still indicates $r_c/r_g \sim 3$ only. Anyway, this model cannot represent a realistic cosmological scenario, having too much power on large scales. Thus, it seems that the basic scenarios do not fully reproduce the observed effect.

A trend with richness can be qualitatively understood in terms of statistics of rare events in a linear Gaussian noise. Kaiser[92] showed that ξ for mass in high density regions, above a given threshold, increases with the height of the threshold, indicating a trend with richness for ξ_c. Such an idealized model, however, cannot make realistic quantitative predictions, especially in pancake scenarios, mainly because initial density peaks are not uniquely related to Abell-like clusters; many peaks end up in pancakes or filaments, so that some degree of local isotropy is required to guarantee an Abell-like cluster, and nonlinear effects may also be important. In $n = -2$ models,[93] r_c derived from linear density peaks was found to overestimate the one derived for realistic clusters by a factor of 2 or more, indicating that the statistical amplification of rare events cannot explain the whole observed effect even in such models.

A hint may come from indications that clusters tend to populate elongated or flattened structures[7)-9)] of $\lambda_c \sim 100 - 150\text{h}^{-1}$Mpc, perhaps to be called "superpancakes" in analogy with the "pancakes" of galaxies of $\lambda_g \sim 20 - 30\text{h}^{-1}$Mpc. I showed in § 2 that the correlation function near unity is strongly affected by the *dimensionality* of the distribution of points [Eqs. (17)-(18)], and that pancaking which results from a coherence length at λ_g can indeed by responsible for the observed excess of ξ_g in the range $2 - 10h^{-1}$Mpc. An analogous geometrical effect may produce the observed excess of ξ_c on larger scales: an *additional coherence length* at $\lambda_c \simeq 5\lambda_g$ would produce $\sim 100 - 150\text{h}^{-1}$Mpc superpancakes of small amplitude, inducing preferential cluster formation along them, yielding $r_c \simeq 5r_g$. The present "linear" rms amplitude of perturbations on the pancake scale should be $\delta(\lambda_g) \simeq 1.5$ in order to fit the appropriate slope of ξ_g (§ 3). The amplitude on the superpancake scale should be large enough to affect the formation of clusters, but small enough not to affect ξ_g too much; $\delta(\lambda_c) \sim 0.5$ is appropriate. This process is demonstrated in Fig. 11 by a kinematic N-body simulation[93] in which the points represent galaxies and clusters are identified at peaks where a local 3D collapse is anticipated, *i.e.*, where the lowest eigenvalue of the deformation tensor [Eq. (9)] is above some critical positive value chosen to give the appropriate number density for $R \geq 1$ Abell clusters. Figure 12 shows the correlation functions from these simulations, with $r_c/r_g \simeq 5$ as required. The bias towards cluster formation in superpancakes is enhanced if the universe is open, as marginally bound lumps end up as clusters while marginally unbound lumps freeze out in the expanding universe.[94]

My conclusion is that a second coherence length at $\lambda_c \sim 100 - 150h^{-1}$Mpc can help explaining the excess of cluster-cluster clustering. I can see two reasonable possibilities for the *cosmological origin* of such a feature.; the *photon-baryon Jeans length* prior to decoupling [Eq. (1)], or the neutrino damping length [Eq. (3)], both if the universe is open with $\Omega \sim 0.1$.

7. Conclusions: The Dark Matter

In the competition between the cosmological scenarios I ($n = 0$), A (coherence length, dissipative pancake), AI (hybrid, nondissipative pancake), and the "cold" particles, I would summarize the current scores as follows:

(1) The correlation function near unity argues against the I scenario, and for pancakes that have formed recently. This is incompatible with the A scenario and is a strong point in favor of the AI scenario. The "cold" spectrum provides the appropriate correlation length if $\Omega h \sim 0.2$, or if the galaxies formed selectively in regions of some high density. Tentative evidence for recent evolution in $\xi(r)$, based on Lyman-alpha clouds and deep galaxy surveys, argues against scenarios I and A, and against selective galaxy formation.

(2) The presence of "pancakes" of $\sim 30h^{-1}$Mpc, as indicated by the alignment of clusters and superclusters, and more qualitatively by other tests like the percolation test or by the presence of large voids, provides another evidence against the I scenario, and perhaps the strongest current evidence against the "cold" particle scenario with $\Omega = 1$. It argues in favor of the pancake scenarios, and can be reproduced by cold particles as well if $\Omega h < 0.2$.

(3) The null detection of galaxy-pancake alignment provides an argument against the A scenario. The simple fact that galaxies exist far away from pancakes is another strong argument against the A scenario.

(4) The excess of cluster-cluster correlation indicates nonvanishing power on very large scales. This argues either for a features in the spectrum at a very large scale, or for a very flat spectrum ($n < -1$) on very large scales.

The "mid-term" grades based on the above tests would be: "fail" for the I ($n = 0$) scenario, "poor" for the A scenario, "fair" for the "cold" particles, and "good" for the AI scenario, but the study is still far from final conclusions.

Based on the above conclusions, I would argue that the three basic structures — galaxies (including subgalactic structure and small clusters), pancakes (including rich clusters, strings, and voids), and superpancakes (if indeed distinguishable from pancakes) — arise from different features in the linear spectrum of density perturbations at the beginning of the gravitational instability era. These should be an I component that dominates on scales below a few Mpc, an A component with a coherence length at $\lambda_g \sim 30h^{-1}$Mpc, and a large scale coherence length at $\lambda_c \sim 150h^{-1}$Mpc (I would regard this last requirement to be more tentative). The constraints posed by these requirements on the possible nature of the dark matter and its contribution to the universal density will be summarized below.

The dark matter should meet other major cosmological constraints as well. Although inflational models prefer $\Omega = 1$ (if $\Lambda = 0$), the astronomical dynamical evidence point to $\Omega \sim 0.1$ for the material that participates in clustering on all scales. The Hubble constant is constrained to be in the range $0.5 \leq h < 1$, and the ages of globular clusters, which seem to be older than $12\,Gyr$ (conservatively!) indeed require $h \sim 0.5$ and $\Omega \leq 1$. Estimates based on element abundances and primordial nucleosynthesis[95)] constrain the baryonic contribution to be $\Omega_b \sim 0.1$, and although there are ways around this constraint,[96)] I will not consider a larger value for Ω_b here. Finally, the observational upper limits on small scale anisotropies in the microwave background temperature,[97)] $\delta T/T$, if they survive from the original last scattering at $z \sim 10^3$, rule out pure baryonic perturbations and "cold" particles with $\Omega \leq 0.2$. However, reionization of the universe at $10 < z < 100$ can smear out $\delta T/T$ on angles below 7°, and release this constraint.

TABLE 1: Large scale constraints on the dark matter

Structure:			Galaxies	Pancakes	Superpancake
Linear spectrum:			I on small scales	$\lambda_g \sim 30h^{-1}$ Mpc	$\lambda_c \sim 150h^{-1}$ Mpc
Dark Matter					
Ω_b	Ω_ν	Ω_c			
0.1			Pop III	λ_s	λ_J
0.1	1		–	λ_ν	–
0.1	0.1		Pop III	λ_s	λ_J or λ_ν
0.1		1	"cold"	–	–
0.1		0.1	"cold" or Pop III	λ_s, "cold"	λ_J

The ability of the different dark matter candidates to provide the required features in the perturbation spectrum is summarized in Table 1. It turns out that an Einstein-de Sitter universe is not very successful. On one hand, a universe dominated by massive neutrinos ($\sim 30eV$; $\Omega_\nu \sim 1$) provides pancakes and voids in a natural way via the damping length at λ_ν, but fails to form galaxies in time. On the other hand, a universe dominated by "cold" particles ($\Omega_c \sim 1$) gives rise to galaxies very naturally, but fails to produce large pancakes and voids. Both fail to explain superpancakes. It seems that we are forced to overcome some theoretical prejudice and consider an *open* universe.

If $\Omega \sim 0.1$, the baryonic contribution to the density is important even if neutrinos and/or "cold" particles are present. If the perturbations are adiabatic, the Silk damping scale would provide pancakes and voids, and the photon-baryon Jeans length would provide superpancakes. Also, the cold particles would help producing pancakes on the same scale, and λ_ν would help producing superpancakes. Galaxies may arise from isothermal perturbations of baryons or from a "cold" particle spectrum. This I component of subgalactic perturbations can form Population III objects[98)] which end up as baryonic dark matter in the form of small unevolved stars (Jupiters), or compact objects (black holes, neutron stars, or white dwarfs). These Population III objects can naturally provide the required energy for reionizing large volumes before $z \sim 30$, and maintaining the ionization long enough to ensure large optical depth which will isotropize the microwave background radiation. I conclude that a pure *baryonic* dark matter can naturally provide the required features in the perturbation spectrum which results in a satisfactory scenario for the formation of large scale structures.

The approach in the above analysis was to study the formation of structure in the recent gravitational instability free of assumptions based on models of the early universe and the question of the origin of the perturbations. The conclusions call for an open mind concerning prejudices that are fashionable among cosmologists. the universe is more likely to be open, with $\Omega \sim 0.1$ today, and it can very well be dominated by the conventional baryonic dark matter. If one assumes $\Omega = 1$ after all, the more exotic scenarios, based on massive neutrinos or "cold" particles, are in trouble. The origin of an appropriate perturbation spectrum made of two components (AI) has to be explored further.

I gratefully acknowledge my collaborators S. J. Aarseth, and M. J. West. This material is based upon research supported in part by the NSF Grant No. PHY77-27084, supplemented by funds from NASA.

References

1. Bahcall, N. 1977, Ann. Rev. Astr. Ap. 15, 505.
2. Oort, J.H. 1983, Ann. Rev. Astr. Ap. 21, 373.
3. de Vaucouleurs, G. 1978, in The Large Scale Structure of the Universe, ed. M.S. Longai and J. Einasto (Dordrecht: Reidel), p. 205.
4. Tully, R.B. 1982, Ap. J. 257, 389.
5. Yahil, A., Sandage, A., and Tammann, G. 1980, Ap. J. 242, 448.
6. Bahcall, N. and Soneira, R. 1984, Ap. J. 277, 27.
7. Ciardullo, R., Ford, H., Bartko, F., and Harms, R. 1984, preprint (Space Telescope 7).
8. Giovanelli, R. and Haynes, M.P. 1982, A.J. 87, 1355.
9. Batuski, D.J. and Burns, J.O. 1984, preprint (submitted to M.N.R.A.S.).
10. Einasto, J, Joeveer, M., and Saar, E. 1980, Nature 283, 47.
11. Tago, E., Einasto, J., and Saar, E. 1984, M.N.R.A.S. 206, 559.
12. Ostriker, J.P. and Cowie, L. 1981, Ap. J. (Lett.) 243, L127.
13. Primack, J.R. and Blumenthal, G.R. 1984, in Formation and Evolution of Galaxies and Large Structures in the Universe, ed. J. Audouze, and J. Tran Thanh Van (Dordrecht: Reidel), p. 163.; Primack, J.R. this volume.
14. Zeldovich, Ya.B. 1970, Astr. Ap. 5, 84.
15. Peebles, P.J.E. 1980, The Large Scale Structure of the Universe (Princeton: Princeton University Press).
16. Aarseth, S.J. 1984, in Methods of Computational Physics, ed. J.U. Brackbill and B.I. Cohen (New York: Academic Press), p. 1.
17. Efstathiou, G., Davis, M. Frenk, C.S., and White, S.D.M. 1984, preprint (ITP).
18. Dekel, A. 1981, Ap. J., 264, 373.
19. Dekel, A. and Aarseth, S.J. 1984, Ap. J. 283, August 1.
20. Gott, J.R., III, and Rees, M.J. 1975, Astr. Ap. 45, 365.
21. Hogan, C.J. and Kaiser, N. 1983, Ap. J. 274, 7.
22. Silk, J. 1968, Ap. J. 151, 459.
23. Peebles, P.J.E. and Yu, J.T. 1970, Ap. J. 162, 815.
24. Silk, J. and Wilson, M. 1980, Phys. Scripta 21, 708.
25. Wilson, M.L. and Silk, J. 1981, Ap. J. 243, 14.
26. Peebles, P.J.E. 1981, Ap. J. 248, 885.
27. Bonometto, S.A., Caldara, A., and Lucchin, F. 1983, Astr. Ap. 126, 377.
28. Wilson, M.L. 1983, Ap. J. 273, 2.
29. Jones, B.J.T. and Wyse, R.F.G. 1983, M.N.R.A.S. 205, 983.
30. Bond, J.R., Efstathiou, G., and Silk, J. 1980, Phys. Rev. Lett. 45, 1980.
31. Doroshkevich, A.G., Khlopov, M.Yu., Sunyaev, R.A., Szalay, A.S., and Zeldovich, Ya.B. 1981, Ann. NY Acad. Sci., 375, 32.
32. Bond, J.R. and Szalay, A.S. 1983, Ap. J. 274, 443.; Szalay, A.S., this volume.
33. Peebles, P.J.E. 1982, Ap. J. (Lett.) 263, L1.
34. Peebles, P.J.E. and Dicke, R.H. 1968, Ap. J. 154, 892.
35. Peebles, P.J.E. 1984, Ap. J. 277, 470.
36. Doroshkevich, A.G. 1970, Astrofizika 6, 581.
37. Lin, C.C., Mestel, L., and Shu, F.H. 1965, Ap. J. 142, 1431.
38. Zeldovich, Ya.B., Einasto, J., and Shandarin, S.F. 1982, Nature 300, 407.
39. Sunyaev, R.A. and Zeldovich, Ya.B. 1972, Astr. Ap. 20, 189.
40. Dekel, A. 1983, in Early Evolution of the Universe and Its Present Structure, ed. G. Abell and G. Chincarini (Dordrecht: Reidel), p. 429.
41. Aarseth, S.J., Gott, J.R., III, and Turner, E.L. 1979, Ap. J. 228, 664.
42. Davis, M., Efstathiou, G., Frenk, C.S., and White, S.D.M. 1984, in preparation.
43. Davis, M. and Peebles, P.J.E. 1983, Ap. J. 267, 465.
44. Einasto, J., Klypin, A.A., Saar, E., and Shandarin, S.F. 1984, M.N.R.A.S. 206, 529.
45. Davis, M. and Peebles, 1977, Ap. J. Suppl. 34. 425.
46. Peebles, P.J.E. 1974, Ap. J. (Lett.) 189, L51.
47. Efstathiou, G. and Eastwood, J.W. 1981, M.N.R.A.S. 194, 503.
48. Frenk, C.S., White, S.D.M., and Davis, M. 1983, Ap. J. 271, 417.

49. Klypin, A.A. and Shandarin, S.F. 1983, M.N.R.A.S. 204, 891.
50. White, S.D.M., Frenk, C.S., and Davis, M. 1983, Ap. J. (Lett.) 274, L1.
51. Centrella, J. and Melott, A. 1983, Nature 305, 196.
52. Dekel, A. and Shapiro, P.S. 1984, in preparation.
53. Kaiser, N. 1984, in Inner Space/Outer Space, ed. E.W. Kolb and M.S. Turner (University of Chicago Press).
54. Bardeen, J. 1984, in Inner Space/Outer Space, ed. E.W. Kolb and M.S. Turner (University of Chicago Press).
55. Koo, D.C. and Szalay, A.S. 1984, Ap. J. 282, in press.
56. Tyson, A. 1982, private communication.
57. Sargent, W.L.W., Young, P.J., Boksenberg, A., and Tytler, D. 1980, Ap. J. Suppl. 42, 41.
58. Sargent, W.L.W., Young, P.J., and Schneider, D.P. 1982, Ap. J. 256, 374.
59. Weymann, R. 1984, preprint.
60. Dekel, A., West, M.J., and Aarseth, S.J. 1984, A. J. 279, 1.
61. Carter, D. and Metcalfe, N. 1980, M.N.R.A.S. 191, 325.
62. Binggeli, B. 1982, Astr. Ap. 107, 338.
63. Binney, J. and Silk, J. 1979, M.N.R.A.S. 188, 273.
64. Palmer, P.L. 1983, M.N.R.A.S. 202, 561.
65. Einasto, J., Klypin, A.A., Saar, E., and Shandarin, S.F. 1984, M.N.R.A.S. 206, 529.
66. Aarseth, S.J. and Saslaw, W.C. 1982, Ap. J. (Lett.) 258, L7.
67. Dekel, A. and West, M.J. 1984, in preparation.
68. Kuhn, J.R., and Uson, J.M. 1982, Ap. J. (Lett.) 263, L47.
69. Moody, J.E., Turner, E.L., and Gott, J.R. 1980, Ap. J. 273, 16.
70. Vishniac, E. 1984, in preparation.
71. Fesenko, B.I. 1982, Soviet Astr. Lett. 8, 247.
72. Doroshkevich, A.G., Kotok, E.V., Shandarin, S.F., Sigov, Yu.S. 1983, M.N.R.A.S. 202, 537.
73. Roth, J. and Djorgovski, S. 1984, preprint.
74. Chernoff, D. and Davis, M. 1984, private communication.
75. Shandarin, S.F. and Zeldovich, Ya.B. 1982, Comments Ap. 10, 33.
76. Dekel, A. and West, M.J. 1984, Ap. J., in press.
77. Bhavsar, S.P. and Barrow, J.P. 1983, M.N.R.A.S. 205, 61.
78. Melott, A.L., Einasto, J., Saar, E., Suisalu, I., Klypin, A.A., and Shandarin, S.F. 1984, Phys. Rev. Lett. 51, 935.
79. Thompson, L.A. 1976, Ap. J. 209, 22.
80. Strom, S.E. and Strom, K.M. 1978, Astron. J. 83, 732.
81. Adams, M.T., Strom, K.M., and Strom, S.E. 1980, Ap. J. 238, 445.
82. Djorgovski, S. 1983, Ap. J. (Lett.) 274, L7.
83. Helou, G. and Salpeter, E.E. 1982, Ap. J. 252, 75.
84. MacGillivray, H.T., Dodd, R.J., McNally, B.V., and Corrin, H.G. 1982, M.N.R.A.S. 198, 605.
85. Kaprandis, S. and Sullivan, W.T. 1983, Astr. Ap. 118, 33.
86. Flin, P. and Godlowski, W. 1984, in Clusters and Groups of Galaxies, ed. F. Mardirossian, G. Giuricin, and M. Mezzetti (Dordrecht: Reidel).
87. Dekel, A. 1984, preprint (ITP).
88. Bahcall, N. and Soneira, R. 1983, Ap. J. 270, 20.
89. Schectman, S.A. 1984, preprint.
90. Barnes, J., Dekel, A., Efstathiou, G., and Frenk, C. 1984, preprint (ITP).
91. Shandarin, S.F. and Klypin, A.A. 1984, preprint.
92. Kaiser, N. 1984, Ap. J. (Lett.), in press.
93. Dekel, A. 1984, Ap. J. 284, Sept. 15.
94. Dekel, A. 1981, Astr. Ap. 101, 79.
95. Yang, J., Turner, M., Schramm, D., Steigman, G., and Olive, K. 1984, Ap. J. 281, 493.
96. Rees, M.J., private communication.
97. Uson, J.M. and Wilkinson, D.T. 1984, Ap. J. (Lett.) 277, L1.
98. Carr, B.J. 1984, in Inner Space/Outer Space, ed. E.W. Kolb and M.S. Turner (University of Chicago Press).

49. [illegible] A. and [illegible] J. P. 1984, M.N.R.A.S. 209, 801.
50. Mould, J. [illegible] C. [illegible] and Davis, M. 198[illegible] Ap. J. (Lett.) [illegible]
51. [illegible] and [illegible] A. [illegible] 1983 Nature 305 [illegible]
52. [illegible] A. and Shapiro, P. [illegible] in preparation.
53. [illegible] A. 1984 in Inner Space/Outer Space, ed. E. W. Kolb and M. S. Turner (University of Chicago Press).
54. [illegible] 1984 in Inner Space/Outer Space, ed. E. W. Kolb and M. S. Turner (University of Chicago Press).
55. [illegible] 1984, Ap. J. 28[illegible] in press.
56. [illegible] 1984, private communication.
57. [illegible] Babul, A. and Taylor, D. 1980, Ap. J. [illegible]
58. [illegible] Schramm, D. N. 1982, Ap. J. 276 [illegible]
59. [illegible]
60. [illegible] 198[illegible]
61. [illegible]
62. [illegible]
63. [illegible]
64. [illegible]
65. [illegible]
66. [illegible]
67. [illegible]
68. [illegible]
69. [illegible] 276, 46[illegible]
70. [illegible] in press.
71. [illegible] 24[illegible]
72. [illegible] Ap. J. 1984 [illegible]
73. [illegible]
74. [illegible]
75. [illegible]
76. [illegible]
77. [illegible]
78. [illegible]
79. Thompson, L.A. [illegible] Ap. J. 2[illegible]
80. [illegible]
81. [illegible] 1986, Ap. J. 28[illegible]
82. [illegible] 1983, Ap. J. (Lett.) 274, [illegible]
83. [illegible] 1982, Ap. J. [illegible]
84. [illegible] 1982, M.N.R.A.S. [illegible]
85. [illegible] and [illegible] 1982, Phil. [illegible] 179 [illegible]
86. [illegible] 1984 in Clusters and Groups of Galaxies, ed. F. Mardirossian, [illegible] and M. Mezzetti (Dordrecht: Reidel).
87. Dekel, A. 1984, preprint (ITP).
88. [illegible] and [illegible] 1984, Ap. J. 276, [illegible]
89. [illegible] A. 1984, preprint.
90. [illegible] Dekel, A., [illegible] and [illegible] 1984, preprint (ITP).
91. [illegible] and Primack, J. A. 1984, preprint.
92. [illegible] 1984, Ap. J. (Lett.), in press.
93. [illegible] A. 1984, Ap. J. 281, [illegible]
94. [illegible] A. [illegible] Ap. J. 10[illegible]
95. [illegible] Ostriker, J. P. [illegible] and [illegible] 1984, Ap. J. 281, 493.
96. [illegible] M. [illegible] private communication.
97. [illegible] and Wilkinson, D.T. 1984, Ap. J. (Lett.) 277, L1.
98. [illegible] 1984 in Inner Space/Outer Space, ed. E. W. Kolb and M. S. Turner (University of Chicago Press).

WHERE DO GALAXIES FORM ?

ALEXANDER S. SZALAY

Institute for Theoretical Physics
University of California
Santa Barbara, California 93106
and
Department of Atomic Physics
R. Eötvös University
Budapest, Hungary

ABSTRACT

The present theories of galaxy formation are reviewed. The relation between peculiar velocities and the correlation function of galaxies points to the possibility that galaxies do not form uniformly everywhere. None of the present theories of galaxy formation can account for this fact in a natural way. Theoretical implications of several observations as Lyman-α clouds, correlations of faint galaxies are discussed.

1. Initial Conditions

The universe contains a wide dynamic range of objects : from stars (1 M_0) all the way to superclusters (10^{16} M_0). A major question that we are unable to answer yet is whether the formation of structure has started with smaller masses clustering on ever larger scales[1)], or whether extremely large structures formed first, then subsequently fragmented into smaller ones[2)]. If we knew the precise initial conditions then the present structure of the universe could be derived by applying the laws of physics. Let us summarize, what has to be known about the initial conditions for this ambitious project.

- Type of fluctuations

  The fluctuations are likely to be adiabatic, since the specific entropy of the universe, n_B/n_γ is tied to microscopic parameters of particle physics. Entropy fluctuations, once popular, can only be generated by huge amounts of shear.

- Origin of fluctuations

  In the inflationary theories quantum fluctuations arise in a natural way. However, the necessary amplitude seems to require rather special prescriptions for the effective potential[3)].

- Spectrum of fluctuations

  The initial perturbations are expected to be scale free. therefore their Fourier amplitude depending on the wavenumber k can be well described by a power law, $|\delta_k|^2 \propto k^n$. If the spectral index is $n = 1$, the amplitude of the different perturbations is the same when their wavelength equals the horizon size. This 'double scale-invariant' is called the Zeldovich spectrum, and is known to arise in inflationary scenarios[4].

Amplitude of fluctuations/ limits on $\Delta T/T$

There are severe constraints on the fluctuation amplitudes. If the fluctuations were adiabatic, the perturbations of the metrics generate fluctuations in the temperature of the microwave background. On small angular scales (4.5 arc mins) these limits are extremely small[5]:

$$\Delta T/T < 2.9 \times 10^{-5} \quad .$$

The standard growth of fluctuations in a flat universe is $(1+Z)^{-1}$. The H-He plasma becomes gravitationally unstable only after recombination, at $Z \sim 1000$. At this point the density and temperature fluctuations are similar, $3\ \Delta T/T = \Delta\rho/\rho$. This does not leave enough margin for fluctuation growth, the fluctuations cannot reach the nonlinear stage our universe seems to be in today. Present calculations confirm[6,7] that if the universe is baryon dominated, only prohibitively high initial fluctuation amplitudes can result in the formation of galaxies. If the universe is dominated by some form of collisionless dark matter, the dark matter fluctuations are unaffected by pressure, therefore grow even before recombination. After recombination these curvature perturbations caused by the dark matter will accelerate fluctuation growth in the baryons, so the $\Delta T/T$ constraints are less stringent.

Dark matter / processes modifying the initial spectra

Though the initial spectrum is a power law, by the time it becomes nonlinear it will be considerably modified. When the universe is radiation-dominated, fluctuations within the horizon have a minimal increase[8], whereas the ones outside the horizon grow. This effect will bend the slope of the spectrum from n to $n-4$ for wavenumbers higher than k_{eq}, corresponding to the size of the horizon when the matter and radiation energy densities were equal. The presence of the collisionless dark matter results in distortions of a different kind: the free motion of particles erases structures smaller than the free streaming scale[9,10,11,12]. The mass scale of this collisionless damping process can be expressed in terms of the mass and entropy of the particles the dark matter is consisting of.

$$M_x \approx 2.2\ m_p^3 m_x^{-2}$$

In the case of neutrinos this mass takes the value of $M_{\nu m} = 3.2 \times 10^{15} m_{30}^{-2}\ M_{\odot}$, corresponding to the comoving length scale $\lambda_{\nu m} = 41\ m_{30}^{-1}$ Mpc. Depending on what the 'temperature' of the dark matter is, this damping scale can change from the above 41 Mpc to extremely small values. The neutrinos are *hot* particles, since their average momentum is close to that of the background radiation photons. Most other candidates for the dark matter like axions and photinos - yet undiscovered - would have decoupled much before the neutrinos, having a lower entropy or temperature, so they are called *cold*. They hardly move at all, their damping scale is negligible. Intermediate candidates, like a gravitino of 1 keV mass would be *warm*.

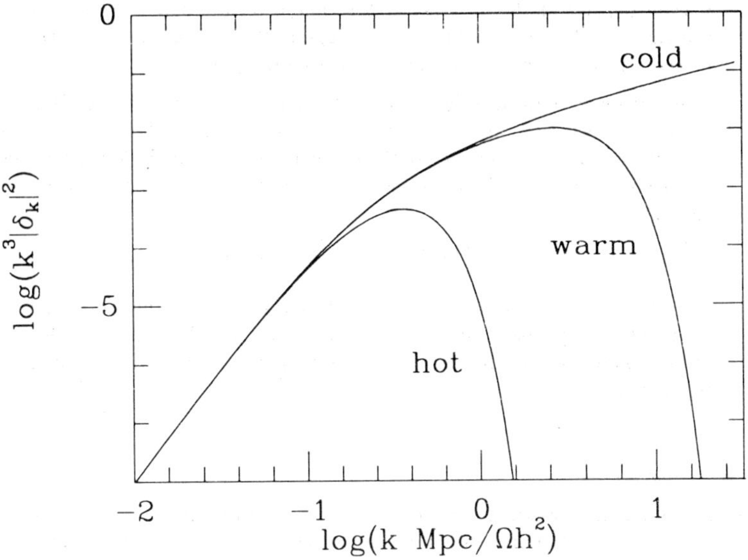

Fig.1. The shape of the mass fluctuation spectrum $log(\Delta M/M) = log(k^3\ |\delta_k|^2)$ as a function of wavenumber k for different candidates of the dark matter[9).]

- Phase correlations/ Gaussian random noise

  A major underlying assumption in calculating most consequences of a given fluctuation spectrum is that the phases of the individual Fourier components are random, ie. the perturbations are a random Gaussian process. One can envisage scenarios, where this will not be the case, like perturbations originating from strings[13)] For a given spectrum combined with the assumption of random phases one can calculate the distribution of mass fluctuations, density of local peaks, density profiles around local peaks, the distribution of peaks of a given size, etc.

  If galaxy formation is a Gaussian process, there is no sudden onset, galaxies form at an ever increasing rate. Since today at $Z = 0$ there is very little sign of galaxy formation, this process was inhibited at some point, though the mechanism is not clear yet.

- Global properties of the universe

  The expansion of the universe is characterized by three quantities: $\Omega = \rho/\rho_{crit}$, the density parameter, H_0, the Hubble constant, Λ_0, the cosmological constant. $\Omega = 1$ corresponds to the flat universe, which appears to be necessary for inflation. Λ_0 is generally assumed to be negligable. However, $\Omega = 1$ and $H_0 = 50$ km/s.Mpc with $\Lambda_0 = 0$ imply uncomfortably low values for the present age of the universe. Calculations of the primordial ^{4}He and D+^{3}He abundance indicate[14)], that the baryon density of the universe at the time of primordial nucleosynthesis lies in the range of $0.01 < \Omega_B < 0.1$. This suggests that if baryons dominate the mass density then the universe is open by a large margin.

Fluctuation growth also depends on the density of the universe. If $\Omega < 1$, the growth of perturbations effectively stops at the redshift $Z = \Omega^{-1}$. The detailed predictions of $\Delta T/T$ are below the current limits if the dark matter consists of neutrinos with about 30 eV mass, but restrict Ω if the cold particles dominate the universe [6,7].

$$\Omega \geq 0.2 \times h^{-4/3} \approx 0.5 \times h_{50}^{-4/3}$$

$h = H_0/100$ km/s.Mpc and $h_{50} = H_0/50$ km/s.Mpc, dimensionless. It was assumed that galaxies follow the mass distribution: the amplitude of the fluctuations today was normalized to J_3, the integral of the galaxy-galaxy correlation function $\xi(r)$.

2. Nonlinear structure

Here we would like discuss the expected structure of the universe if the dark matter is either hot, warm or cold. Once the first mass scale in a spectrum with a large damping cutoff (hot) reaches nonlinearity, particle trajectories cease expanding away from each other and converge, resulting in the temporary formation of caustics. The density becomes very high and a flat 'pancake' is formed[2]. At first they arise at isolated spots where the initial velocity perturbations had the largest gradient. Soon these regions grow, turning into huge surfaces which intersect, forming the walls of a cell-structure which is itself gravitationally unstable.

In this nonlinear phase mode-mode coupling among Fourier components sends power to short wavelengths, and correlates the phases even though the initial fluctuation spectrum may have had random ones. The methods of catastrophe theory were applied[15] to analyze structure that develops in such potential motion. It was found that the two dimensional pancakes are only the lowest order singularities; other singular topological structures should also appear. String-like features are one example, and they can be seen in the N-body simulations.

When the intersection of trajectories takes place, gas pressure builds up, the velocity of the collapsing gas exceeds the sound speed and a shock wave is formed[2]. The gas is shock-heated up to keV temperatures and cools by emitting radiation over a broad spectrum. Recently several authors[16] have calculated the cooling of collapsing neutrino-baryon pancakes, the details of which are considerably different from those in a pure baryon pancake[17]: the baryon density is lower, infall velocities are higher, thus the cooling rate is much slower. It is evident that the fraction that can cool significantly is a sensitive function of the mass of the collapsing region. This cooling is necessary, since only cold gas is able to form the seeds of galaxies, so the local column density *modulates the rate of galaxy formation.*

The UV and soft X-ray emission can photoionize the intergalactic medium, making galaxy formation in regions that have not yet formed pancakes more difficult, which would accentuate the contrast in galaxy density between the strings and pancakes vs. voids, even though the density contrast may be only 3-10.

If the dark matter is warm, it will still form pancakes, though of galactic size. There the cooling is much more efficient[18], those timescales will determine the fate of each object. If the dark matter is cold, the spectrum $k^{3/2}\ \delta_k$ is substantially different from the hot and warm case. It has no peak at all, but it is slowly increasing towards the smallest scales. These small scales will collapse first, but later the larger systems are also going nonlinear, forming a clustering hierarchy. Due to the complicated nature of these many-body interactions only numerical N-body simulations are able to follow the evolution of such systems.

3. N-body simulations

If we had the answers to all the parameters listed above, it would be relatively easy to calculate the evolution of the universe. On collisonless dark matter only gravitational forces are acting, so one can numerically solve the transport equations, even in the nonlinear regime. This has indeed been done, as we discuss here. Given the initial conditions, these numerical experiments will tell us the mass distribution in the universe. One can hope, that the structure obtained this way will resemble the real universe, ie. galaxies trace the mass distribution.

Starting from the above mentioned initial conditions extensive N-body simulations consisting of more than 32000 particles [19,20] were made. These projects all used some version of a particle/mesh Fourier code, 64^3 in size. The calculations were started, when $\delta\rho/\rho$ was about 0.2, and the approximate Zeldovich solution[2] corresponding to the growing mode of perturbations was used to determine initial positions and velocities, then the trajectories of the particles were integrated. The free parameters of the calculations are Ω, H_0 and the initial amplitude of the fluctuations. For a given Ω one can use conservative limits for the age of the universe to obtain a value of H_0. If $\Omega = 1$, then $t_0 > 12$ Gy requires $H_0 < 54$ km/s.Mpc.

The initial amplitude can be defined in different ways. For simulations with hot dark matter, neutrinos, the epoch of galaxy formation, Z_{GF} was defined as the redshift when 1 percent of all particles have gone through a 'caustic'. One can see on Fig.2a the evolution of the mass autocorrelation for this case (Davis, Frenk, Efsthathiou and White[19]), and unless $Z_{GF} \approx 0.4$, the correlation function disagrees with that of the galaxies.

For cold dark matter the initial amplitude is determined in a different way. Fig.2b indicates the evolution of the mass autocorrelation function for cold dark matter. Due to the growth of nonlinearity, $\xi(r)$ is rapidly increasing both in slope and amplitude, just like for hot dark matter. One can define *today* when the correlation function of the particles most resembles that of the galaxies, ie. a power law with a slope -1.8.

$$\xi(r) = (r/r_0)^{-1.8} \ .$$

However, at this point the amplitude is too small. One can resolve this difficulty by choosing $H_0 = 22$ km/s.Mps, but this is hardly the way out.

There is one more difficulty: the random velocity dispersion of galaxies is well known[21]:

$$< v_{12}^2 >^{1/2} \approx 300 - 400 \mathrm{km/s}.$$

In the neutrino simulations, if $Z_{GF} > 1$ the corresponding velocity dispersions are in the 1200 km/s range, clearly too high. In linear perturbation theory

$$< v_{12}^2 > \approx (180\ \mathrm{km/s})^2\ f(\Omega)\ \xi(0)(H_0\lambda_{\nu m})^2.$$

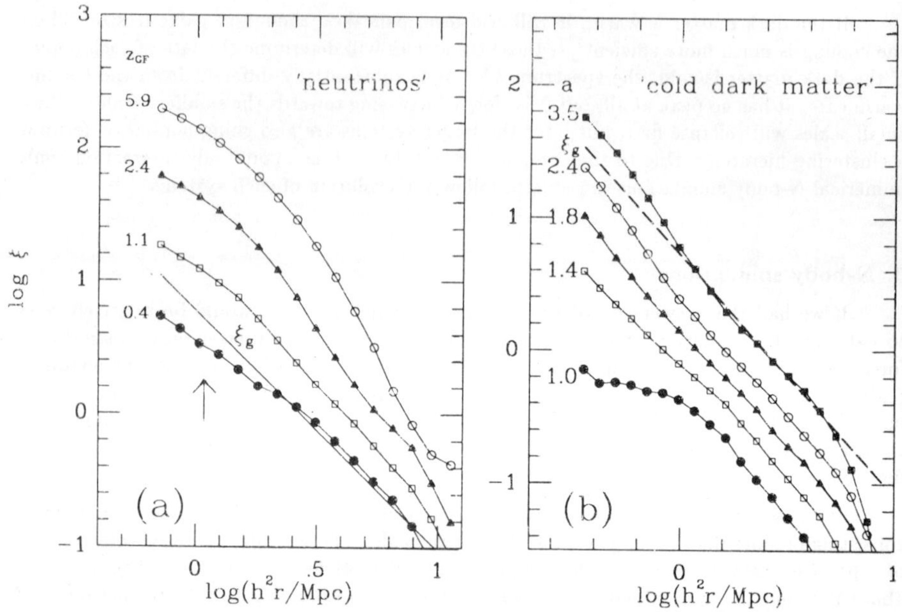

Fig.2. The evolution of the mass autocorrelation $\xi(r)$ in the simulations of Davis, Frenk, Efsthathiou and White[19] for neutrino dominated (a) and cold dark matter dominated (b) universe.

Today, $\xi(0) \approx (1 + Z_{GF})^2$, so either $\Omega << 1$, forbidden by the $\Delta T/T$ constraints, or Z_{GF} is small[22].

For cold dark matter a similar problem exists, and since a low Ω model is ruled out, the only remaining possibility is to have $\xi(0) = |\delta\rho/\rho|^2$ fairly small. Then we are in an even sharper contradiction with the observed galaxy autocorrelation.

On the other hand, the galaxies consist mostly of baryonic gas capable of emitting and absorbing radiation. These dissipative processes, strongly density and temperature dependent, occur at a different rate at different places[18]. All these effects, combined with possible shock waves due to the finite pressure in the H-He gas, may have an important role in determining where galaxies form. As a result, the galaxies may not follow the light at all, so the mass autocorrelation should not be compared to the galaxy autocorrelation. Galaxy formation, as long as it is a random process, initiated by gravitational infall will be likely to start at the regions of highest densities. One can therefore associate the particles in these regions with galaxies. This 'biasing' of galaxy formation towards these high densities is a heuristic procedure, but probably a fair approximation to what really happens. The detailed numerical procedure for selecting these 'biased' particles in the cold dark matter models involved a smoothing of the densities before the actual selection was made. The physical explanation of where the threshold of the selection should be is much less clear, it can only be adjusted to the observed number density of galaxies. This 'biasing' process will enhance the correlations, as can be seen on Fig.3a,b without having large peculiar velocities. This enhancement makes

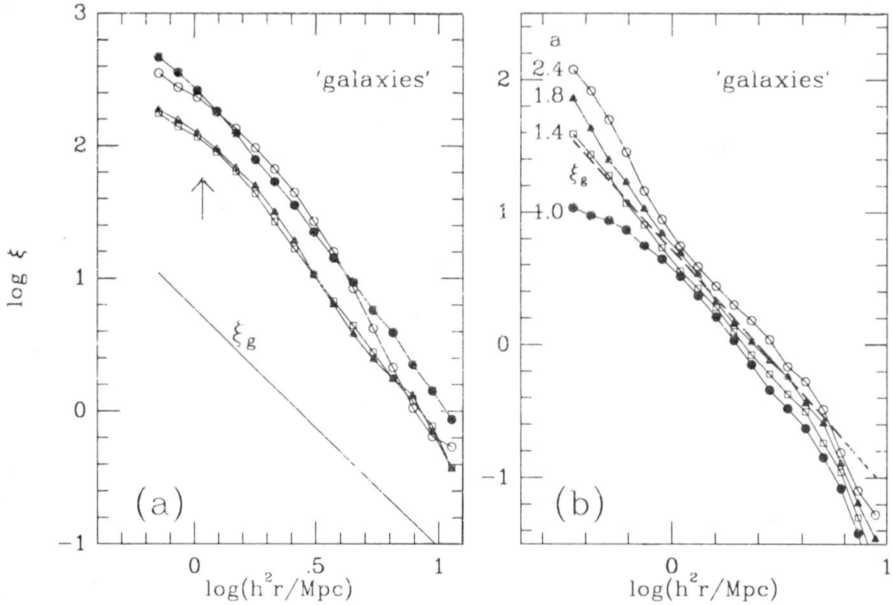

Fig.3. The evolution of $\xi(r)$ for 'galaxies' in the neutrino dominated (a) and the cold dark matter dominated (b) simulations [19]. Particles at the highest density regions were selected as galaxies.

the cold particles actually work, as far as the agreement with ξ_g is concerned.

In the neutrino dominated universe the particles which have crossed the caustics at any given time are the ones associated with 'galaxies'. This selection will also dramatically increase the correlation, making the model incompatible with observations. The particles selected this way contain all the regions, where galaxies may form, but not all these particles can be galaxies. Most of the contribution to the small scale correlation is coming from the regions where tight clumps are. If the rate of galaxy formation could be biased against those regions, the correlation could be somewhat decreased.

4. Observations

Here we would like to discuss observations of relevance to galaxy formation, even if light does not trace the mass. Some of these problems are known for some time, but they were not considered as 'fundamental' tests of galaxy formation theories, whereas they may be.

- Ly-α absorption systems

In the spectra of high redshift QSO's several Ly-α absorption systems were found[23]. The typical characteristics of this Lyα-forest are narrow (10-40 km/s) absorption lines with typical neutral H column densities of $N_{HI} \approx 10^{13-14}$ cm^{-2}. The line width sets an upper limit to the temperature of the clouds, $T_{cloud} < 30000\ K$. The neutral H number density in the clouds is about $10^{-3.5-4.5}$, about a 10 times overdensity at those redshifts. There are about 40 clouds in a unit redshift interval, between redshifts 2 and 3. This corresponds to a mean separation along the line of sight to more than 100 Mpc, but this is $< nR^2\pi >^{-1}$, the mean free path, not the real mean separation $< n >^{-1/3}$. The cloud sizes are about 10-30 kpc[24], which tells us that the comoving number density of clouds has to be 100 to 1000 times higher than that of bright galaxies. These clouds appear to be unclustered, $\xi_{clouds} < 0.1\xi_g$.

If the universe is dominated by cold dark matter, this provides another clue, that we need 'biased' galaxy formation. If galaxies, when formed, were evenly distributed just as the Lyα clouds are, and their present correlation is due to their gravitational motion, one would expect the same correlation for the clouds. The clouds could not be destroyed sufficiently when near to a galaxy to explain the lack of correlations. If the galaxies are clustered anomalously 10 times stronger, then the clouds have just the ordinary clustering properties.

If the universe is neutrino-dominated, the hot gas in the pancakes has extremely high temperature and pressure, so the clouds cannot exist in those regions, the external pressure would compress them. The only place, where they could survive, would be in the big voids. This would explain their low correlation, but it seems to be extremely hard to generate the required small scale fluctuations within the voids.

The formation process of the Lyα clouds is still not clear, and the major uncertainty is the state of the intergalactic medium, its temperature and density. It can be shock heated, but then the release of the energy driving the shocks must be associated with explosions related to galaxy formation processes, or it can be photoionized. In the latter case the source of the necessary photons is somewhat unclear.

- Stability of $\xi(r)$

From Fig.2a,b it can be seen, that the correlation function of the mass is evolving rather rapidly both in slope and amplitude. At the redshift 0.5 one can see significant differences. From angular correlations $w(\theta)$ of very faint J=24 galaxies limits were obtained[25] how ξ behaves at that redshift, and these limits are clearly incompatible with the results of the simulations. Since ξ for the 'biased' galaxies (Fig.3a,b) is extremely stable, this points again to biased galaxy formation.

- Redshift surveys, filaments and voids

Redshift surveys seem to be the best way to determine the real distribution of galaxies in 3 dimensions. The first such surveys caused a lot of excitement, because they indicated, that galaxies are not uniformly distributed over space, but rather they occupy a few percent of the available volume. They are often found in long chain-like filaments, like the Perseus Supercluster. They leave large voids, like the Bootes, about 60 Mpc in diameter behind. Such structures can easily arise in the neutrino dominated picture , but they are hard to form in with cold dark matter. Presently there is not enough data to assess the statistical significance of the presence of the voids.

- QSO distribution

Recently the luminosity function of quasars was determined using extremely faint observations[26]. The data indicate, that the comoving density is *decreasing* at higher redshifts. This is another feature, that has to be accounted for in the galaxy formation scenarios.

- cluster-cluster correlations

It has been known for some time, that Abell clusters have a much larger correlation length, than galaxies do[27)], but it was realized only recently, how hard it is to explain this feature. The theory of rare Gaussian events[28)], higher order correlation functions[29)] are recent attempts for explanation.

5. Conclusion

All the present theories of galaxy formation fail to explain the observed universe in its full complication. The recent experimental development on the microwave background fluctuations provides the strongest constraints on the present theories. The details of the galaxy correlation properties are a new challenge, indicating that galaxies are unlikely tracers of the mass distribution.

6. Acknowledgements

I would like acknowledge useful discussions with Simon White, Dick Bond, Nick Kaiser and David Koo. Figures 2. and 3. are results from recent calculations by Davis, Efstathiou, Frenk and White.

7. References 1. Peebles,P.J.E. 1980, The Large Scale Structure of the Universe, Princeton University Press, Princeton, N.J.

Zeldovich,Ya.B. 1970, Astron.Astrophys. 5, 84.; Sunyaev,R.A. and Zeldovich,Ya.B. 1972, Astron.Astrophys. 20,189.

Guth,A. and Pi,S-Y. 1982, Phys.Rev.Lett. 49, 1110.; Steinhart,P. and Turner,M.S. 1984, Phys.Rev. D29, 2162.; Brandenberger,R. this volume.

Brandenberger,R. and Kahn,R. 1984, Phys.Rev. D29, 2172.; Bardeen,J., Steinhart,P. and Turner,M.S. 1983, Phys.Rev. D28, 679.; Brandenberger,R. 1984, Rev.Mod.Phys. in press.

Uson,J.M. and Wilkinson,D.T. 1984, Ap.J.Lett. 277, 1.; Uson,J.M. and Wilkinson,D.T. 1984, Proc. Inner Space/Outer Space Workshop, Fermilab, in press.

Bond,J.R. and Efsthathiou,G.P. 1984, Proc. Inner Space/Outer Space Workshop, Fermilab, in press.

Vittorio,N. and Silk,J. 1984, Ap.J.Lett. in press.

Guyot,M. and Zeldovich,Ya.B. 1970, Astr.Ap. 9, 227.; Meszaros,P. 1974, Astr.Ap. 37, 225.

Gershtein,S.S. and Zeldovich,Ya.B. 1966, JETP Lett. 4,174.; Cowsik,R. and McClelland,J. 1972, Phys.Rev.Lett. 29,669.; Marx,G. and Szalay,A.S. 1972, Proc Neutrino '72, 1,123.; Szalay,A.S. and Marx,G. 1976, Astron.Astrophys. 49,437.; Schramm,D.N. and Steigman,G. 1981, Ap.J. 243,1.; Bond,J.R., Efsthathiou,G. and Silk,J. 1980, Phys.Rev.Lett. 45,1980.; Doroshkevich,A.G., Khlopov,M.Yu., Sunyaev,R.A., Szalay,A.S. and Zeldovich,Ya.B. 1980a, Ann. N.Y. Acad.Sci. 375,32.; Doroshkevich,A.G. and Khlopov,M.Yu. 1981, Sov. Astron. 25,521.

Bond,J.R. and Szalay,A.S. 1981, Proc. Neutrino 81, 1, 59.; Bond,J.R. and Szalay,A.S. 1983, Ap.J., 274, 443.; Bond,J.R., Szalay,A.S. and Turner,M.S. 1982, Phys. Rev. Lett. 48, 1636.;

11. Peebles,P.J.E. 1982, Ap.J. 258, 415.; Peebles,P.J.E. 1982. Ap.J.Lett. 263, 1.
12. Pagels,H. and Primack,J. 1982, Phys.Rev.Lett. 48, 223.; Blumenthal,G.R., Pagels,H. and Primack,J. 1982, Nature 299, 37.; Blumenthal,G.R., Faber,S.M., Primack,J.R. and Rees,M.J. 1984, Nature, in press.
13. A.Vilenkin 1984, Proc. Inner Space/Outer Space Workshop, Fermilab, in press.
14. Olive,K.A., Schramm,D.N., Steigman,G., Turner,M.S. and Yang,J. 1981, Ap.J. 246,557.
15. Arnold,V.I., Shandarin,S.F. and Zeldovich,Ya.B. 1982, Geophys.Astrophys. Fluid Dynamic 20,111.
16. Bond,J.R., Centrella,J., Szalay,A.S. and Wilson,J.R. 1984, M.N.R.A.S. to be published. Shapiro,P.R., Struck-Marcell,C. and Melott,A. 1983, Ap.J. 275, 413.
17. Doroshkevich,A.G., Shandarin,S.F. and Saar,E. 1978, M.N.R.A.S. 184,643.
18. Rees,M.J. and Ostriker,J. 1977, M.N.R.A.S. 179,541.; Silk,J. 1977, Ap.J. 211,638.
19. Davis,M., Efsthathiou,G.P., Frenk,C.S. and White,S.D.M. 1984, submitted to Ap.J.; Frenk C.S., White, S.D.M. and Davis, M. 1983, Ap.J. 271, 417.; White,S.D.M. 1984, Proc. Inne Space/Outer Space Workshop, Fermilab, in press.h
20. Centrella,J. and Melott,A. 1983. Nature 305, 196.; Klypin,A.A. and Shandarin,S.F. 1983 M.N.R.A.S. 204, 891.; Melott,A. 1983, M.N.R.A.S. 202,595.
21. Davis,M. and Peebles,P.J.E. 1983, Ap.J. 267, 465.
22. Kaiser,N. 1983. Ap.J.Lett. 273, 1.
23. Young,P., Sargent,W.L.W. and Boksenberg,A. 1982, Ap.J. 252, 10.
24. Foltz,C.B., Weymann,R.J., Röser,H.J. and Chaffee,F.H. 1984, Ap.J.Lett. 281, 1.
25. Koo,D.C. and Szalay,A.S. 1984, Ap.J. 282, in press.
26. Koo,D.C. and Kron,R. 1984, personal communication; Koo,D.C. 1983, Proc. "Quasars and Grav. Lenses", Liege Coll. p.240.
27. Peebles,P.J.E. and Hauser,M.G. 1974, Ap.J.Suppl. 28, 19.; Bahcall,N.A. and Soneira,R.M 1982, Ap.J.Lett. 258, 17.
28. Kaiser,N. 1984, Proc. Inner Space/Outer Space Workshop, Fermilab, in press.
29. Fry,J.N. 1984, Proc. Inner Space/Outer Space Workshop, Fermilab, in press.

PROBLEMS OF DIMENSIONAL REDUCTION AND INFLATIONARY COSMOLOGY*

David Appel and Max Dresden

Institute for Theoretical Physics
State University of New York
Stony Brook, New York 11794

ABSTRACT

We discuss various aspects of the relationship between Kaluza-Klein theories and cosmology, with particular emphasis on the process of dimensional reduction. Some features of higher-dimensional Friedmann-Robertson-Walker-type universes are discussed, such as the compatibility of the Einstein equations, their connection with inflationary scenarios, and their consequences for the critical density of the observable four-dimensional universe. The role of the Casimir effect as a possible mechanism for dimensional compactification is explored. The absence of the Casimir effect for supersymmetric theories is noted. Various subtleties of the Casimir effect make it difficult to decide at this point whether the Casimir effect is the operative mechanism in dimensional reduction.

*This is an expanded and streamlined version of a talk given by M. Dresden.

Outline

(1) Some general observations and an old concern.
(2) The compatability of extra dimensions and inflation.
(3) The Casimir effect as the mechanism for dimensional compactification.
(4) Conclusions, questions and some new concerns.

(1) Some General Observations and An Old Concern

a. The extra dimensions

In the last few years, there has been a substantial and increasing interest in the study of cosmologies of universes with more than four dimensions. There are a number of distinct reasons for this dramatic upswing in the interest of physicists in spaces possessing extra dimensions. The earliest efforts in this direction, the Kaluza[1)]-Klein[2)] theories were unquestionably motivated by the desire to construct unified theories of electrodynamics and gravitation. In the Klein version of the theory, the resulting quantization of the electric charge was an additional attractive feature of a five-dimensional gravitational theory. Although no rigorous proofs exists, it now appears possible and even plausible that extra-dimensional theories might provide a suitable framework for unified theories of strong, electroweak and gravitational interactions. Such theories would be distinct alternatives to the more conventional grand unified schemes (SU(5) and up). This possibility (or this hope) was and is a major motivation in the construction of many-dimensional theories[3)]. Another stimulus for the growing interest in higher-dimensional theories is provided by supergravity. Again, no complete rigorous deductions exist but there are a large number of scattered results which all suggest that supergravity in either 10 or 11 dimensions possesses a number of special formal properties which make these spaces especially suitable for superunified theories.

Actually, the idea that geometrical features of the spacetime would be connected with--or even define--physical characteristics is quite old; it was certainly a significant element in Einstein's general philosophy. There is no very precise or rigorous formulation of this idea, but the general expectation is that the extra dimensions (more than 4) are in some way connected with internal quantum numbers, such as strangeness or color. Only in one case has this connection between an internal quantum number and geometrical properties been unequivocally established. The spin was first introduced as an "intrinsic" or internal angular momentum associated with rotations of electrons. It was later recognized that the "spin" is completely described as a Casimir operator of the Poincaré group. Thus, the "internal" or "model" spin quantum number was replaced by a purely group theoretical notion. The program of the "geometrization of physics" is to make this type of connection into a systemic and general procedure. It is still interesting to recall that Pais[4], faced with the problem of providing a geometrical (or group theoretical) basis for an internal quantum number (the strangeness) remarked as early as 1953 that the 4-dimensional Poincaré group could not accommodate additional quantum numbers, that groups with more Casimir operators would be needed. These ideas eventually (and in a somewhat roundabout way) led to the exploration of internal symmetry groups. In this process, the direct association of geometrical notions in higher-dimensional spaces with physical concepts was largely lost. At that time, for example, no serious effort was made to connect specific physical concepts with many-dimensional ($d>4$) geometrical concepts. In the current thinking, this particular connection is taken up again. The underlying belief is that the extra dimensions do have something to do with the physical world; that there are indeed phenomena or effects which are remnants or subtle consequences of the existence of these extra dimensions. They are not envisaged as merely formal bookkeeping devices, but their existence should have physical consequences.

b. Cosmological Aspects of the Extra Dimension

Although one may contemplate the possibility of extra spacetime dimensions, the fact remains that in the phenomena of particle physics, there is no hint or indication that additional spacetime dimensions are necessary or useful. However, at sufficiently high energies, these extra dimensions can be expected to play an important role; in particular, this could be true in the very early universe. All later stages (after 10^{-10} seconds after the big bang) are well described by more or less conventional physics in 4 dimensions. In the course of time, the extra dimensions, their influence and effect, must have diminished or changed. The precise manner in which this process took place is not so clear. An easy way to visualize this process is that in the course of time, the usual three dimensions kept on expanding, while the extra dimensions, very rapidly (within 10^{-38}-10^{-40} sec.?) contracted to distances of the order of the Planck length. Conceivably, the internal degrees of freedom are remnants or manifestations of the extra dimensions. This contraction process is usually referred to as "dimensional reduction" or "dimensional compactification". A very interesting consequence of this type of evolutionary process is that it has recently been shown that in the process of dimensional reduction, a considerable amount of entropy is produced[5,6]. In turn, the rapid production of a large amount of entropy in the universe allows, in principle, an explanation of the horizon and flatness problems[5,6,7], different from that provided by the inflationary universe (new or old). This indicates that at least some aspect of the dimensional reduction process lead to physical consequences.

Taking the extra dimensions seriously leads to the conclusion that in the course of time, some kind of a contraction or compactification process must have taken place.

This observation leads to two classes of problems: compatibility questions and questions concerning the dynamical causes of the compactification mechanism. This paper is concerned with a preliminary exploration of these questions.

c. The Compatibility Question

The time evolution of the universe is--in the usual theory described by the Einstein equations--in 1 + 3 dimensions. It is well known that these equations allow solutions in which the 3 space dimensions expand in the course of time. When considering a universe with extra dimensions, a question of compatibility comes up: Do the Einstein equations, in N dimensions, allow solutions where d dimensions expand and D contract in the course of time? Here N=1+d+D; d is the number of usual spatial dimensions; there is one single time dimension. Since the left-hand side of the Einstein equations is purely geometrical, $(R_{AB} - \frac{1}{2} g_{AB} R)$, its generalization to N dimensions is unambiguous and unique. The right hand side of the Einstein equations, the energy-momentum tensor of the matter T_{AB}, under the assumptions customarily made in cosmology, also allows a unique generalization to N dimensions. Consequently, it is reasonably straightforward to investigate the temporal behavior of the Einstein equation in N dimensions. It is then possible to check whether the compactification type process is a possible stage in the developmental sequence. Such studies have been carried out by a number of authors for N=5, as well as for general N. Some of these studies are numerical[8,7]; in the present paper, section (2) is devoted to a qualitative study of these questions. All studies agree in the basic points: the Einstein equations allow solutions whose temporal development combines the expansion of some dimensions with the compaction of other dimensions. It is, however, not correct that the Einstein equations demand that particular form of the solution. It appears that the Einstein equations allow but do not require dimensional compactification.

d. The Casimir Effect As the Cause of Dimensional Compactification

Even though it is interesting that the Einstein equations allow compactification, this by itself gives no hint as to why this dimensional reduction would occur. The other group of questions, strongly suggested by the idea of compactification, is to find reasons within the now known physics, which would cause compactifications to occur. Thus rather than merely using a type of temporal evolution allowed by the Einstein equations, one looks for reasons which would require that particular evolutionary process to occur. Classical thermodynamics and classical phase transitions provide an example of the type of argument envisaged. In classical thermodynamics, the formal connections are distinct from the stability considerations. The stability condition that the total free energy of a total system is a minimum, requires that under appropriate conditions, the system splits into two distinct phases. This splitting is, of course, allowed by formal (differential) thermodynamic laws, but they do not require it. This splitting into two disjoint phases becomes mandatory only when the stability conditions are invoked. One might speculate that an analogous situation might exist in the compactification problem. Some principle or law might prefer or demand a space which is split into two disjoint subspaces, a D-dimensional compact or bounded space and a (d+1)-dimensional non-compact (unbounded) space. The compactification itself would then appear as a "phase transition of the geometry itself" with the unbounded, non-compact space playing the role of the unbounded gas phase, and the bounded compact space would be the counterpart of the finite volume liquid phase. At the present time, there appears to be only one known physical mechanism which might accomplish such a reduction and that mechanism is provided by the Casimir effect. The Casimir energy--the zero-point energy of a confined quantum field--can be positive or negative, depending on the nature of the field, the geometry, and the topology of the confining space. As was first explicitly calculated by Casimir[9], this zero-point energy gives rise to forces on the boundaries. It has been proposed that a Casimir-type effect is responsible for dimensional compactification. If correct, this would make the compactification a quantum effect. The quantum field, confined to a particular geometry, possesses a particular ground state energy. Because of this, forces are exerted on the boundaries. These are the forces driving the compactification. Presumably, the final result of the operation of these forces would be the selection of a particular geometry and topology. If the conjecture that the Casimir effect is the mechanism of dimensional reduction is anywhere near correct, the Casimir stresses might act in such a way that compactified subspaces are energetically favorable. In section (3), some aspects of the Casimir effect as they pertain to dimensional reduction will be summarized. Of special importance is the role of the supersymmetric Casimir effect. One of the motivations for the serious considerations of extra-dimensional theories is the simplicity (and uniqueness!) of supergravity theory in 11 dimensions. Witten[10] pointed out an almost miraculous coincidence. He noted that 7 dimensions is the minimum dimensionality a manifold must possess in order to support an SU(3) x SU(2) x U(1) symmetry. Thus a theory in which the SU(3) x SU(2) x U(1) symmetric gauge fields arise as components of the gravitationsl fields in more than 4 dimensions must at least possess 7 extra dimensions, leading to a space of 4+7=11 dimensions. Interestingly enough, the maximum number of dimensions for a supergravity theory in which no particles of spin larger than 2 occur, is eleven. It is therefore important to recognize that the Casimir effect, proposed as the mechanism for dimensional reduction, strictly vanishes in an exact supersymmetric theory. A proof of this important result will be outlined in section

(4). The result might appear to rule out the Casimir effect as the cause of the dimensional reduction. Unfortunately, at high temperatures, especially those at the very early stages of the universe, the Casimir effect becomes small (although the supersymmetry breaking is large). Fortunately, the Casimir effect becomes large for small size confining regions. Thus, the net effect of these competing factors is really a quantitative matter and, consequently, not easy to survey. But the possibility that the Casimir effect is indeed the compactification mechanism is certainly not excluded by present information and the present numbers. Some further comments in these matters will be made in the final section (4).

e. A Concern: The Assumption of Thermal Equilibrium

The compactification process is clearly a time-dependent process. It has been emphasized[11] that almost all the cosmological discussions assume that the system considered (the universe) is and remains in thermal equilibrium at each instant of time. Practically all treatments assume either an equilibrium state, or a state so near equilibrium that a spatially homogenous linearized Boltzmann equation will give an adequate description of the system. This, of course, is not evident, and may not even be true. As always, whether an equilibrium description is legitimate depends on the relative characteristic time scales. In the evolutionary process of the universe, there are a number of such scales: the duration of collisions, the time between collisions, the expansion rate (a la Hubble), the inflationary expansion rate of some dimensions, or the contraction rate of others. It is not at all easy to decide whether the conditions necessary for the establishment and maintenance of thermal equilibrium pertain in the various stages of cosmological evolution. One might be especially concerned about the inflationary phase, the reheating phase, or the dimensional contraction itself. The relevance of this comment is that in a non-equilibrium configuration, the temperature notion itself may not make much sense. One can, of course, always define a temperature at each spacetime point. But the trouble is that in non-equilibrium situations, this quantity fluctuates so wildly that its average value gives little information about the actual behavior of the system. Although the assumption of thermal equilibrium as such is certainly questionable, it has the great advantage that it leads to tangible specific results. Not assuming equilibrium, while logically possible and from a rigorous deductive viewpoint even preferable, has unfortunately not led to any results so far. The assumption of equilibrium or "near" equilibrium is necessary to get any concrete results. But sooner or later it should be scrutinized so that its validity may be checked. Until such time, it remains a nagging concern.

f. Inflation

In the last few years, a number of evolutionary scenarios have been proposed[12,13,14] which although differing in detail, all possess the common feature that the universe at perhaps 10^{-35} sec. after the big bang experienced a dramatic, exponential expansion. After perhaps 10^{-33} sec. of this expansion, other processes took place (a phase transition and reheating) and the universe, after this explosive expansion, entered upon a period of the gentler Hubble expansion. There is no necessary logical relation between the (conjectured) dimensional reduction and the (conjectured) inflation. But it is certainly highly suggestive to explore possible relations between these two processes. Both deal directly with the change of geometry of the universe in time. The compatibility question raised in section (1c) is relevant for both processes. Indeed, it will be shown in section (2) that both an inflationary scenario and dimensional reduction are allowed by the Einstein equations. Although the precise connection between dimensional compactificatin and inflation is certainly not clear, it appears that the two phenomena both can be accommodated within the usual Einstein equations and the assumptions customarily made in evolutionary cosmology. It would be somewhat surprising if two processes, both involved with radical temporal changes in the spacetime geometry, would be wholly independent of each other. Consequently, in the succeeding discussion, both will be kept in mind throughout.

(2) The Compatibility Questions

The evolutionary development of the usual four dimensional theory is obtained from the Einstein equations, together with a number of symmetry assumptions. These well known formulae are assembled here just to set the notation.

a. The Usual Formalism

i. The conditions of spatial isotropy and homogeneity lead to the Friedmann-Robertson-Walker form of the metric:

$$(ds)^2 = -(dt)^2 + R^2(t)\left[\frac{dr^2}{1-kr^2} + r^2 d\Omega^2\right] \tag{1}$$

where, as usual, k=0 for a flat space; k=+1 for a closed space; k=-1 for an open space. R(t) specifies the "expansion rate" of the universe.

ii. The "cosmological symmetry principles" lead to severe restrictions on the forms of the possible tensors in the spaces. Specifically, in a maximally symmetric subspace, with metric $g_{\alpha\beta}$, the any form-invariant second rank tensor $T_{\alpha\beta}$ is necessarily proportional to the metric tensor:

$$T_{\alpha\beta} = c\, g_{\alpha\beta}\,. \tag{2}$$

In (2), c cannot depend on the coordinates of the subspace. (For a discussion and proof of these results, see Weinberg[15]). The Robertson-Walker space (1), splits up into two maximally symmetric subspaces, a one-dimensional space, parametrized by the the time t, and a 3-dimensional space parametrized by x^1, x^2, x^3 (always denoted by x^i). Using this maximally symmetric character, it is shown (for example, in reference 15) that any tensor $T_{\mu\nu}$ in that space has the form (μ=0,1,2,3; i=1,2,3)

$$T_{00} = \rho(t) \qquad T_{i0} = 0 \qquad T_{ij} = p(t)\, g_{ij} \,. \tag{3}$$

The more customary and more elegant form of the energy-momentum tensor in that space can then be expressed as:

$$T_{\mu\nu} = (p+\rho)\, U_\mu U_\nu + p\, g_{\mu\nu} \tag{4}$$

where

$$U^0 = 1 \qquad U^i = 0 \,. \tag{5}$$

iii. The combination of (4), (1) and the Einstein equations leads to the well-known evolution equations:

$$\ddot{R} = -\frac{4\pi}{3} G (\rho + 3p) R \tag{6}$$

$$H^2 + \frac{K}{R^2} = \frac{8\pi G}{3}\rho \,, \quad H \equiv \frac{\dot{R}}{R} \,. \tag{7}$$

The only point in repeating these well-known relations is to stress once again that although the symmetry arguments alone lead to the ideal fluid form of the energy-momentum tensor (4), these arguments are not sufficient to demonstrate that ρ and p, as defined by (3), are indeed the physical energy density and physical pressure. For example, from the symmetry arguments alone, it can not be inferred that there should be any relation at all connecting p and ρ. The assumption of an equation of state linking p and ρ, to a common variable, the temperature, depends crucially on the physical picture and, as such, on the physical interpretation attributed to ρ and p.

iv. It is a consequence of the vanishing of the covariant

derivative of $T^{\mu\nu}$, that

$$\dot{\rho} + 3(p+\rho)\frac{\dot{R}}{R} = 0 \tag{8}$$

or

$$\frac{d\rho}{dR} = -\frac{3(p+\rho)}{R}\ . \tag{8a}$$

The specific evolutionary process obtained from (6),(7),(8) depends critically on the relationship between p and ρ. For example, the case p=0, corresponds to a universe filled with cosmic dust, p = ρ/3 is a radiation-filled universe. Each leads to a characteristic temporal development. The inflationary scenario emerges immediately if one assumes p=-ρ. In that case, (8) shows that ρ is constant, say ρ_0 and then (6) leads directly to an exponential expansion for R(t):

$$R(t) = R_0\, e^{t\sqrt{\frac{8\pi}{3}G\rho_0}}\ . \tag{9}$$

In the next section (6), the procedure outlined here in steps i,ii,iii,iv, will be applied to a space with extra dimensions.

b. The General Formalism

i. The problem to be treated in this section is the evolutionary development of a universe in N=1+d+D dimensions described by the Einstein equations. It is assumed that the N-dimensional spacetime manifold can be decomposed into three distinct submanifolds: time, ordinary space (M_d) and a compact space (M_D). Each of these submanifolds is assumed to be homogeneous and isotropic, so each one is maximally symmetric. Because of these assumptions, the metric tensor can be written in the form:

$$g_{AB} = \mathrm{diag}\left[-1,\ R^2(t)\tilde{g}_{ij}(x^k),\ a^2(t)\tilde{g}_{MN}(y^P)\right]. \tag{10}$$

R(t) and a(t) are the scale factors for the d- and D-dimensional spaces respectively. The indices A,B,...run from 0 to N-1; i,j,k...run from 1 to d; μ,ν...from 0 to d, M,N,P...from d+1 to d+D. $\tilde{g}_{ij}$ and $\tilde{g}_{MN}$ are the metric tensors of the maximally symmetric d- and D-dimensional subspaces. Occasionally, n=d+D=N-1, the total number of spatial dimensions, is used. It is straightforward to compute the connection coefficients and the Ricci tensor for this space (see Freund[16]). The nonvanishing components of the Ricci tensor are:

$$R_{00} = d\frac{\ddot{R}}{R} + D\frac{\ddot{a}}{a},$$

(11a)

$$R_{ij} = -g_{ij}\left[\frac{2k}{R^2} + \frac{d}{dt}\left(\frac{\dot{R}}{R}\right) + \left(d\left(\frac{\dot{R}}{R}\right) + D\frac{\dot{a}}{a}\right)\frac{\dot{R}}{R}\right],$$

(11b)

$$R_{MN} = -g_{MN}\left[\frac{2k'}{a^2} + \frac{d}{dt}\left(\frac{\dot{a}}{a}\right) + \left(d\frac{\dot{R}}{R} + D\frac{\dot{a}}{a}\right)\frac{\dot{a}}{a}\right].$$

(11c)

k and k' are the curvature constants of the d- and D-dimensional submanifolds. The Einstein equations in the N-dimensional space are:

$$R_{AB} - \frac{1}{2}g_{AB}R = -8\pi\bar{G}\,T_{AB}. \tag{12}$$

$\bar{G}$ is the gravitational constant appropriate to N dimensions and is given by[22)]

$$\bar{G} = G V_c. \tag{13}$$

V_c is the volume of the compact D-dimensional space. The Einstein equations (12) can be rewritten as

$$R_{AB} = -8\pi\bar{G}\,S_{AB}, \tag{14a}$$

$$S_{AB} = T_{AB} - \frac{g_{AB}}{n-1}T^C_C. \tag{14b}$$

ii. The next step is to construct the energy-momentum tensor T_{AB} for this N-dimensional space. As in the four-dimensional case (see the argument in section a, formula (3)), the required symmetries impose severe restrictions on the form of T_{AB}. The assumed direct sum structure of the metric (10) combined with the assumed maximal symmetry of the subspaces M_d and M_D lead to a special form of the energy-momentum tensor given by

$$T_{AB} = \mathrm{diag}\left[\rho(t),\ p(t)g_{ij},\ p'(t)g_{MN}\right]. \tag{15a}$$

The arguments leading to this form are identical with those leading to the form $T_{\mu\nu}$ in step ii of the previous section. (Strictly speaking, these arguments by themselves would just lead to the more general form:

$$T_{AB} = \mathrm{diag.}\left[\rho(t),\ p(y_M,t)\,g_{ij}\ ,\ p'(x_i,t)\,g_{MN}\right]. \tag{15b}$$

In the sequel, the form (15a) will be assumed; this could be justified by imposing additional symmetry conditions on the N-dimensional space. Form (15a) is already sufficiently general to exhibit a variety of evolutionary behaviors, such as dimensional contractions or expansions which are the main concern of this study). Using (15a), the source term (14b) can be directly obtained as:

$$(n-1)\,S_{00} = (n-2)\rho + pd + p'D\,, \tag{16a}$$

$$(n-1)\,S_{ij} = g_{ij}\left(\rho + (D-1)p - Dp'\right), \tag{16b}$$

$$(n-1)\,S_{MN} = g_{MN}\left(\rho - pd + (d-1)p'\right). \tag{16c}$$

By combining (16) with the Einstein equation (14a) and the form of the Ricci tensor (11), there result three differential equations for a and R:

$$d\,\frac{\ddot R}{R} + D\,\frac{\ddot a}{a} = -\,\frac{8\pi\bar G}{n-1}\left[(n-2)\rho + pd + p'D\right], \tag{17a}$$

$$\left[\frac{2k}{R^2} + \frac{d}{dt}\frac{\dot R}{R} + \left(d\frac{\dot R}{R} + D\frac{\dot a}{a}\right)\frac{\dot R}{R}\right] = \frac{8\pi\bar G}{n-1}\left(\rho + (D-1)p - Dp'\right), \tag{17b}$$

$$\left[\frac{2k'}{a^2} + \frac{d}{dt}\left(\frac{\dot a}{a}\right) + \left(d\frac{\dot R}{R} + D\frac{\dot a}{a}\right)\frac{\dot a}{a}\right] = \frac{8\pi\bar G}{n-1}\left(\rho - pd + (d-1)p'\right). \tag{17c}$$

These equations are basic for the further analysis. It is easy to check that, in the case D=0, d=3, equations (17a) and (17b) are equivalent to the 4-dimensional equations (D=0), (6) and (7).

iii. The basic equations have the same structure as (6) and (7) in the 4-dimensional case. In that case, one has two equations for the variables R(t), p(t) and ρ(t). In the usual cosmology, an equation of state is assumed, which relates both p and ρ to the temperature variable. The two equations for T and R, then codetermine the temporal development of the scale factor R and the temperature T in the epoch where the assumed equation of state is valid. In the present case (equations 17), similar arguments can be given. For example, for a matter-dominated universe, p and p' are both much less than ρ, so in both (16) or (17), the p,p' can be neglected. In that case, the system (17) codetermines the time development of R,a, and ρ. Another possibility is to consider a radiation-dominated universe. In that case, both p and p' are proportional to ρ. That again reduces the system (17) to a system of three simultaneous equations in three variables. Sahdev[8] claims to find preferential dimensional reductions. In all these analyses, d and D as well as k and k' are treated as parameters in the theory.

c. Application to the Inflationary Epoch

The case of special interest here is the inflationary epoch. In the spirit of the inflationary approach, the only form of matter considered is a scalar field ϕ, subject to an effective potential $V(\phi)$. The general idea of inflation is that during this stage, the universe supercools below the critical temperature of a phase transition so that the energy-momentum tensor is dominated by the vacuum energy of the effective potential. It is necessary to calculate ρ,p,p' for that case so that they can be used in (17). The Lagrangian density of the field of mass m in the N-dimensional curved space is L:

$$L = -\frac{1}{2}\sqrt{-g}\left[g^{AB}(\partial_A\varphi)(\partial_B\varphi) + m^2\varphi^2\right] - \sqrt{-g}\, V(\varphi). \tag{18}$$

The equations of motion are as usual:

$$(D_A D^A - m^2)\varphi(x) = -V'(\varphi). \tag{19}$$

D_A is the covariant derivative in the N-dimensional space. The energy momentum tensor T_{AB} is computed in the standard manner; it leads to:

$$T_{AB} = -\frac{2}{\sqrt{-g}}\left\{\frac{\partial L}{\partial g^{AB}} - \frac{\partial}{\partial x^C}\left(\frac{\partial L}{\partial g^{AB}_{,C}}\right)\right\} = \varphi_{,A}\varphi_{,B} - g_{AB}\left[\frac{1}{2}\varphi_{,C}\varphi^{,C} + \frac{1}{2}m^2\varphi^2 + V\right]. \tag{20}$$

The most important assumption made at this point is that ϕ depends only on the time. Thus, we consider only configurations where $\nabla\phi=0$. Such spatially homogeneous configurations can be expected to have less energy than those when

$(\nabla\phi)^2>0$, so that one considers the low-lying energy states as the states which primarily govern the compactification or expansion process. It is further convenient (but not of fundamental significance) to neglect the mass term. With these assumptions, (20) exactly assumes the form (15):

$$T_{AB} = \mathrm{diag}\left[\rho(t), p(t)\, g_{ij}, p'(t)\, g_{MN}\right]. \tag{15}$$

with

$$\rho(t) = T_{00} = \tfrac{1}{2}\dot{\varphi}^2 + V(\varphi), \tag{20a}$$

$$p(t) = \tfrac{1}{d} T^i_i = \tfrac{1}{2}\dot{\varphi}^2 - V(\varphi), \tag{20b}$$

$$p'(t) = \tfrac{1}{D} T^M_M = \tfrac{1}{2}\dot{\varphi}^2 - V(\varphi). \tag{20c}$$

It is interesting that the assumption that ϕ is independent of x^i and y^M leads directly to the equality of p and p'. In the inflationary scenario, as the universe supercools, the V term in T_{AB} dominates over all other kinetic terms. Consequently in that phase, one can read off from (20a,b,c) that

$$p'(t) = p(t) \cong -\rho(t). \tag{21}$$

This is really the "matter" condition for inflation; the first equality in (21) is exact as a glance at (20b) and (20c) shows, the second approximate. Just as the conditions $p=-\rho$ in conjunction with (8) showed the inflationary behavior of R, the analogue of (8) which, in this case, is

$$\dot{\rho} + d(p+\rho)\frac{\dot{R}}{R} + D(p'+\rho)\frac{\dot{a}}{a} = 0 \tag{22}$$

((22) is a direct consequence of $D_A T^{AB}=0$) can in conjunction with (21) lead to inflationary behavior for R and a. If (21) is satisfied, (22) shows that indeed $\rho(t)$ is constant, call it $\overline{\rho}_0$. The constancy of the energy density in the evolutionary process is typical for the inflationary process. It is now straightforward to rewrite the basic equations for the source terms (16) under the conditions (21) $p'=p=-\rho=-\overline{\rho}_0$ which leads to the evolution equations for the inflationary epoch:

$$d\,\frac{\ddot{R}}{R} + D\,\frac{\ddot{a}}{a} = c\,, \tag{23a}$$

$$\frac{\ddot{R}}{R} + (d-1)\left(\frac{\dot{R}}{R}\right)^2 + D\,\frac{\dot{R}}{R}\,\frac{\dot{a}}{a} + \frac{2k}{R^2} = c\,, \tag{23b}$$

$$\frac{\ddot{a}}{a} + (D-1)\left(\frac{\dot{a}}{a}\right)^2 + d\,\frac{\dot{R}}{R}\,\frac{\dot{a}}{a} + \frac{2k'}{a^2} = c\,. \tag{23c}$$

Here, c is a constant given by

$$c = \frac{16\pi \bar{G}\bar{\rho}_0}{n-1} = \frac{16\pi G \rho_0^{(4)}}{n-1}\,. \tag{24}$$

In (24), $\rho_0^{(4)} = \bar{\rho}_0 V_c = \int \bar{\rho}_0 d^D y$. Since $\bar{\rho}_0$ is constant, $\rho_0^{(4)}$ is just the energy density of the N-dimensional space integrated over the compact dimensions. The equations (23) characterize the temporal evolution of the universe under the conditions of inflation (21). It is no doubt clear that the system (23) is a set of three equations for two unknown functions. The conditions for inflation have effectively eliminated one parameter. Consequently, it is no longer obvious that the system (23) is compatible for all d,D,k,k' and c. For that reason, the equations (23) are really compatibility conditions. It is not obvious either that $\dot{a}/a$ and $\dot{R}/R$ necessarily have the opposite signs. A somewhat general analysis of the system (23) would therefore be of considerable interest. This will not be done here, just some comments will be added to further clarify the role of these equations.

i. If one assumes (without much justification!) that both M_d and M_D are flat (k=k'=0), the system (23) decouples. It is easy to check that there are solutions:

$$R(t) = R_0\, e^{Ht}\,, \quad H^2 = \frac{6}{n(n-1)}\, H_I^2\,. \tag{25}$$

In (25), H_I is the usual Hubble constant of the inflationary scenario, given by

$$H_I^2 = \frac{8\pi}{3} G \rho_0^{(4)} . \tag{26}$$

In this same case, one shows directly that

$$\frac{\dot{a}}{a} = \frac{\dot{R}}{R} . \tag{27}$$

Then the submanifolds M_d and M_D inflate or deflate together. This is scarcely surprising, since by picking k and k' both zero the equations (23) are totally symmetrical in a and R; consequently, they have to behave in consort. It would be of some interest in this case and also in general to investigate the initial value problem, but in the symmetrical case (k=k'=0), one can not expect anything else except joint inflation and deflation.

ii. Consider the most interesting non-symmetrical case, with k'=+1 for the D-dimensional space and k=0 for the d-dimensional space. Some manipulations of the equations (23) then yield:

$$2dD\frac{\dot{R}}{R}\frac{\dot{a}}{a} = (n-1)c - \left[d(d-1)\left(\frac{\dot{R}}{R}\right)^2 + D(D-1)\left(\frac{\dot{a}}{a}\right)^2 + \frac{2D}{a^2}\right]. \tag{28}$$

It is clear that the quantity in brackets is always positive. If a is small enough, the product $\frac{\dot{R}}{R}\frac{\dot{a}}{a}$ will become negative, which would cause R and a to go in opposite directions. But the precise behavior can not be obtained from (28) alone; a more careful analysis of the system (23) is needed.

d. <u>The Cosmological Term and a Specific Result</u>

The basic equations (17) are not specialized to the inflationary epoch. They have a number of interesting properties. To mention just one, consider the Einstein equation with D=0, d=3, the ordinary four-dimensional case, but this time with a cosmological term. In that case, the Einstein equation (7) is replaced by:

$$\left(\frac{\dot{R}}{R}\right)^2 + \frac{k}{R^2} = \frac{8\pi G}{3}\rho + \frac{1}{3}\lambda . \tag{29}$$

λ is the cosmological term, which is known to be very small. It is interesting that a very similar equation can be obtained from the system (17) which, as will be recalled, results from the Einstein equation in 1+d+D-dimensions <u>without</u> a cosmological term. For this purpose, put d=3 in (16), but leave D arbitrary, so that n=3+D. Manipulating these equations in this case, one can derive:

$$\frac{\dot{R}^2}{R^2}+\frac{k}{R^2}=\frac{16\pi\bar{G}\rho}{3(D+2)}+\frac{1}{3}D\Big[c'\Big(\frac{3}{2}p-p'\Big)+\frac{\ddot{a}}{a}+\frac{1}{2}(D-1)\Big(\frac{\dot{a}}{a}\Big)^2+\frac{k'}{a^2}\Big]. \tag{30}$$

Equation (30) describes the equation for the variation in scale factor R(t) of the 3-dimensional subspace of the N-dimensional space. It is now interesting to compare (30) with (29). The left-hand sides both describe the time evolution of the scale factor of the non-compact subspaces of the universe. In the case of (29), this is the evolution of the 3-dimensional spatial subspace of a 4-dimensional continuum. The left-hand side of equation (30) describes the identical process; however, the 3-dimensional spatial subspace is now a subspace of a (3+D)-dimensional spatial subspace which, together with the time coordinate, forms a (4+D)-dimensional spacetime. The identity of the left-hand sides of (30) makes it tempting to identify the right-hand sides as well. Because of the minute magnitude of the cosmological term, it is sensible to neglect the λ term in (29) and the $\frac{1}{3}D[\ ]$ term in (30), the first because of observational evidence, the second because of the formal similarity of (29) and (30). It is further important to remember that ρ in (30) is the energy density in the N-dimensional spacetime and shall be written $\rho(N)$, while ρ in (29) is the energy density in the 4-dimensional spacetime and shall be written $\rho(4)$. Thus, (30) can be written as:

$$\frac{\dot{R}^2}{R^2}+\frac{k}{R^2}=\frac{16\pi\bar{G}\rho(N)}{3(D+2)}=\frac{16\pi G V_c\rho(N)}{3(D+2)}. \tag{31}$$

It is natural to define the effective 4-dimensional energy density by:

$$\rho^{(4)}(D+4)=\int d^Dy\,\rho(N)=V_c\,\rho(N). \tag{32}$$

Thus, $\rho^{(4)}(D+4)$ is the average energy density in a 4-dimensional subspace of a space which has D extra dimensions. In terms of $\rho(D+4)$, the basic equation becomes:

$$\frac{\dot{R}^2}{R^2}+\frac{k}{R^2}=\frac{8\pi G}{3}\frac{2}{D+2}\rho^{(4)}(D+4). \tag{33}$$

The condition that the present universe is flat requires k=0; it requires a special value for the Hubble constant $\dot{R}/R$ which is obtained from observations. Thus, for a universe with D extra dimensions, the critical density $\rho_c^{(4)}(4+D)$ is given by:

$$H^2 = \frac{8\pi G}{3}\frac{2}{(D+2)}\rho_c^{(4)}(D+4). \tag{34}$$

On the other hand, if there are no extra dimensions so that equation (29) is operative, the condition for flatness requires the critical density (now called $\rho_c^{(4)}(4)$) to be related to the experimental Hubble constant by:

$$H^2 = \frac{8\pi G}{3}\rho_c^{(4)}(4). \tag{35}$$

Comparing (34) and (35) gives:

$$\rho_c^{(4)}(D+4) = \frac{D+2}{2}\rho_c^{(4)}(4). \tag{36}$$

Thus, if our universe possesses D extra dimensions, the effective critical density is (D+2)/2 times larger than the critical density in a 4-dimensional theory. If we could measure the energy density and independently establish the flatness of the universe, this would in principle be an experimental test of the extra dimensions. If one has 7 extra dimensions, this would be a factor 9/2. It is unlikely that this will be settled in the near future, but it is nevertheless interesting that there are definite numerical effects attributable to the extra dimensions.

3. The Casimir Effect As The Mechanism For Compactification

a. General Comments

As mentioned in section (1d), the Casimir effect is, if not promising, in any case, a possible candidate for the compactification mechanism. It is well-known that the zero-point energy density of a confined quantum field diverges. Nevertheless, Casimir suggested[9] that changes in that zero-point energy would have observable consequences. For photons confined to a rectangular volume of sides L_1, L_2, L_3, the zero-point energy is, of course, $\sum \frac{1}{2}\hbar\omega_k$. With the boundary conditions of a perfect conductor, the energy density is the usual quartically divergent expression:

$$\frac{W_0}{L_1 L_2 L_3} = \frac{1}{2}\hbar c \int d^3k \sqrt{k_1^2 + k_2^2 + k_3^2} \,. \tag{37}$$

(37) is to be understood as a limiting expression, where L_1, L_2, L_3 all approach infinity. If one of the variables is kept fixed and finite, say $L_3 = R_0$, one computes the zero-point energy of the resulting regime $L_1 L_2 R_0$. This again will be infinite. To analyze it, one can introduce a high frequency cut-off Λ in momentum space, (occasionally, $1/\Lambda = \alpha$ will be used). The integrals then become finite. One finds:

$$W_I(\Lambda, R_0) = \hbar \frac{L_1 L_2}{\pi^2} \sum_n \iint dk_1\, dk_2\, \omega e^{-\frac{\omega}{\Lambda}} ;$$

$$\omega = \omega_n(k_1, k_2) = \sqrt{k_1^2 + k_2^2 + \frac{n^2\pi^2}{R_0^2}} \,. \tag{38}$$

Evaluation[17] of (38) gives:

$$W_I(\Lambda, R_0) = \hbar c \frac{L_1 L_2}{\pi^2}\left(3 R_0 \Lambda^4 - \frac{1}{720}\frac{\pi^4}{R_0^3}\right). \tag{39}$$

As the cut-off Λ approaches infinity, W_I is again quartically divergent. One could just subtract out this divergent contribution and obtain the Casimir energy W_C:

$$\frac{W_c}{L_1 L_2} = -\frac{\pi^2}{720}\frac{\hbar c}{R^3} \,. \tag{40}$$

(40) gives rise to attractive forces per unit area:

$$p = -\frac{\partial}{\partial R}\left(\frac{W_c}{L_1 L_2}\right) = -\frac{\hbar c \pi^2}{240 R^4} \,. \tag{41}$$

This, of course, is the well-known Casimir result. The main reason for recalling this derivation is to stress that it requires a renormalization, a regularization. One could either just subtract out the divergent terms or one could more systematically define the Casimir energy by a double limiting process:

$$(\Delta W)_C = \lim_{\alpha\to 0} \lim_{L\to\infty} \left[W_I(R_0,\alpha) + W_{II}(L-R_0,\alpha) - W_{III}(L,\alpha) \right]. \tag{42}$$

The R_0 region is the finite region, $L_1L_2(L-R_0)$. $(L-R_0)$ stands for a region between a finite fixed boundary R_0 and another boundary L which eventually will go to infinity; L is the complete region L_1L_2L. The point to be stressed is that these different regularization schemes, although they all give definite answers, do not always give the same answer. In curved spaces, in particular, serious ambiguities arise in the calculation of Casimir energies. But even in simpler situations (than curved spaces!) the Casimir effects are subtle and often unintuitive. $(\Delta W)_C$, for example, depends on the geometry; the photon Casimir effect is negative (attractive) for plates, positive (repulsive) for spheres. The Casimir energy (even the sign!) depends on the topology of the confining manifold; it is further different for fermi and bose fields. The Casimir effect shows a temperature dependence. For example, the Casimir entropy for a simply connected surface behaves as T^3, while for a multiple connected surface, this entropy varies as T. These unusual and unintuitive features all result from the fact that the actual Casimir calculation always involves nonuniformly convergent expressions; in fact, this nonuniformity is actually the core of the effect. For example, there is a theorem by Hawking which asserts that global Poincaré invariance implies that the energy momentum tensor should be Wick-ordered. This implies (since $\langle 0 | :T_{\mu\nu}: | 0 \rangle = 0$), that there cannot be a Casimir effect. The moment one introduces boundary conditions at a boundary L, the global Poincaré invariance is broken and there can be a Casimir effect. If one now takes the limit $L\to\infty$, the Casimir effect remains, even though formally global Poincaré invariance is restored. This nonuniformity appears to be a general feature. It results physically from the fact (see equation 42) that the effect itself is a finite difference of infinite zero-point energies. Although a number of regularization schemes (ζ function regularization, for example) formulate these subtractions in an ingeneous and even appealing manner, it must be kept in mind that the regularization scheme is a rule adjoined to the physical interpretation. If different rules give different results, as they do in some Casimir calculations, it becomes necessary to find physical reasons which would select one regularization scheme over another.

b. The Casimir Effect and Compactification[18)]

That Casimir-like effects might well play an important role in dimensional reduction was first seriously suggested in an important series of papers by Appelquist and Chodos. In the first of these studies[18)], the authors calculated the influence of quantum fluctuations in a 5-dimensional

Klein-Kaluza theory. They did this by computing the one-loop contritutions to the effective potential. They found after subtracting out a divergent term that the effective potential V^* is given by:

$$V^* = -\frac{15}{4\pi^2}\zeta(5)\frac{\hbar c}{L_5^5}. \tag{43}$$

In (43), L_5 is the circumference of the compact 5th dimension in the Kaluza-Klein theory. L_5 can be expressed as:

$$L_5 = 2\pi R_5 \varphi_c^{1/3}. \tag{44}$$

In (44), R_5 is the radius of the cylindrical part of the Klein-Kaluza manifold. ϕ_c is the classical background field which, for the purposes of calculating the effective potential, is taken to be a constant by Chodos and Appelquist. (43) shows that the effective potential becomes infinitely attractive as $L_5 \to 0$; consequently, it appears that at least in this case, the Casimir effect produces an enthusiastic collapse of the extra dimension. The subtracted term is very similar to the quartically divergent form subtracted in (39), to obtain the finite part of the Casimir effect. Appelquist and Chodos stress that their V^* term is just the gravitational analogue of the Casimir potential. An even more direct demonstration of the "Casimir-like" origin of (43) was given by Rohrlich[20], who directly computed an approximation to the zero-point energy of the gravitational field confined in a Klein-Kaluza manifold. He used the same approximations employed by Appelquist and Chodos writing ($\eta_{\mu\nu}$ is the Minkowski metric):

$$g_{\mu\nu} = \eta_{\mu\nu} + h_{\mu\nu}, \tag{45a}$$

$$\varphi = \varphi_c + \varphi'. \tag{45b}$$

In the calculation, he assumed ϕ_c constant and kept only terms quadratic in ϕ' and $h_{\mu\nu}$. The result of this rather simple straightforward calculation gives for the zero-point energy:

$$W = \frac{15\hbar c}{2\pi^2\alpha^5} + \frac{15\hbar c(2\pi)^3}{L_5^5}\left(5\sum_j \frac{q^2}{(q^2+4\pi^2 j^2)^{7/2}} - \frac{1}{(q^2+4\pi^2 j^2)^{5/2}}\right). \tag{46}$$

In (46), $\alpha = 1/\Lambda$ is the inverse cut-off parameter; it must be taken to the limit $\alpha \to 0$, to obtain a result of physical significance. L_5 is given by (44); q in (46) is defined by:

$$q = \frac{\alpha}{R_5\sqrt{\varphi_c}}. \tag{47}$$

It is important that as $\alpha \to 0, q \to 0$. The expression (46) shows the typical quintic divergence expected for the zero-point energy in a five-dimensional space. If it is subtracted out, one can take the limit $\alpha \to 0$, or $q \to 0$ in (46) and just the terms

$$-\sum_j \left(\frac{1}{4\pi^2 j^2}\right)^{5/2} \tag{48}$$

remain. This gives exactly the ζ function first found by Appelquist and Chodos. It is satisfactory and was to be expected that the Casimir energy can indeed be found by summing the zero-point energy or evaluating the one-loop potential. It is also noteworthy that exactly the same type of divergencies are encountered in both calculations. It should finally be stressed that as carried out, both calculations are approximate. If one carries out a one-loop effective potential calculation by expanding the action around a constant background field, $\phi = \phi_c + \delta\phi$, keeping only quadratic terms in $(\delta\phi)^2$, one obtains exactly the free field Casimir energy. For self-interacting fields, one can still formally write down the zero-point energy but, in that case, higher-loop contributions are needed to reestablish the equivalence between the Casimir energy and the effective potential. In later studies (see especially 19), Appelquist and Chodos do not employ a cut-off procedure with an accompanying subtraction to extract finite results for the Casimir energy. Instead, they use a zeta function renormalization method in which no infinities ever appear. This is indeed an elegant and effective means to obtain finite answers but it must be remembered that no obvious physical significance can be attributed to that regularization procedure.

As has been stressed before, without a physical justification for a regularization, it becomes difficult, especially in curved spaces, to decide which procedure is the most appropriate. It is not at all certain that formal elegance is what is physically desirable. This ambiguity must be kept in mind if the Casimir processes are taken seriously as the mechanisms for compactification. Further investigation[21)] indicates that the mechanisms whereby the Casimir effect affects the dimensional reduction (if at all) is more involved than the computation just outlined. It was shown in reference 20 that in a background manifold $E^d \times (S^1)^D$, (i.e., the direct product of a d-dimensional Euclidean space and a D-dimensional torus), the effective potential due to the Casimir effect is given by:

$$V^*(\ell_1 \cdots \ell_D) = (\cdots) \sum_{m_1=-\infty}^{+\infty} \cdots \sum_{m_D=-\infty}^{+\infty} \frac{1}{(m_1^2 \ell_1^2 + \cdots m_D^2 \ell_D^2)^{\frac{N}{2}}} . \tag{49}$$

$\ell_1 \ldots \ell_D$ are the ranges of the D compact dimensions:

$$0 \leq y_M \leq \ell_M . \tag{50}$$

V^* has no stationary points, so there is really no stable vacuum. Analysis of V^* shows that if one of the D dimensions contracts, the other (D-1) dimensions will expand. This is an example where the Casimir effect, although operative, does not yield a simple picture of dimensional contraction. Strictly speaking, since V^* has no extremes, there appears to be no stable vacuum for that background space. This shows what earlier experiences with the Casimir effect had already indicated: that the size and sign of the effect depends on the topology. At least as serious as the almost capricious dependence on the topology of the background space is the small numerical value of the Casimir effect, especially for matter fields. This is the reason that in reference 23, Candelas and Weinberg needed an enormous number of fields to obtain a reasonable value for their gauge coupling constant. Interesting attempts are now under way to compute the Casimir energy of the gravitational field on a given background geometry (Chodos and Meyers[24)], Chodos[25)]). There is, however, in addition to the ambiguity of the Casimir calculation in curved spaces, the problem that the legitimacy of the one-loop expansion can only be justified a posteriori. It is possible that some of these difficulties--the necessity for a large number of fields, the apparent arbitrariness of the initial topology, the uncertainty about the validity of the one-loop expansion--might be removed or improved in supersymmetric theories. Thus, it is not inappropriate to explore Casimir-like mechanisms in supersymmetric theories and especially in supergravity. There are various reasons for such optimism. In supersymmetric theories, the number, type, and interactions of the fields is fixed and it is not possible (without violating the spirit of supersymmetry) to change their number, let alone introduce thousands of scalar fields. The 11-dimensional supergravity theories also suggest a preferred topology for the underlying space. It is certainly not true that the topology is fixed, but the extant formalism appears to reduce its arbitrariness. Finally, there is some hope (but no certainty) that the regularization schemes of supergravity might be sufficiently general and powerful to overcome many field theoretic ambiguities. Since the ambiguities of the Casimir effect are field theoretic in character, one might hope that in a supersymmetric theory, some of these difficulties might be removed. There is only one trouble with these hopes. There is no Casimir effect in a supersymmetric theory.

c. There is no Supersymmetric Casimir Effect[26)]

i. General

It is very easy to show that supersymmetric theories cannot exhibit a Casimir effect. All supersymmetric theories contain at least one spinor generator Q_α of the supersymmetry transformation. These satisfy

$$\{ Q_\alpha, Q_\beta \} = -(\gamma_\mu C)_{\alpha\beta} P^\mu . \tag{51}$$

C is a charge-conjugation matrix, γ_μ a Dirac matrix, and P^μ the energy-momentum vector. The vacuum state $|0\rangle$ is invariant under supersymmetry transformations, so that

$$e^{iQ_\alpha \lambda_\alpha} |0\rangle = |0\rangle . \tag{52}$$

(52) is true for all (Grassmann) values of λ, thus (52) implies:

$$Q_\alpha |0\rangle = 0 . \tag{53}$$

(51) in the rest system shows that

$$\sum_{\alpha} \varphi_{\alpha}^{2} = (\text{Constants})\, E_0 \tag{54}$$

In (54), E_0 is the energy in the rest system. Combining (54) and (53) shows that the energy of the ground state E_0 satisfies $E_0|0\rangle=0$, hence $E_0=0$; the vacuum state $|0\rangle$ possesses zero energy. The bose and fermi contributions cancel exactly. This appears to eliminate the Casimir effect as a mechanism for dimensional reduction in a supersymmetric theory.

ii. Boundary Conditions--An Example

Actually, the situation is not quite this simple. It has been stressed numerous times that the Casimir effect is a field theoretic effect in a finite region. The essence of a finite region is that the fields must satisfy boundary conditions. These boundary conditions for electromagnetic fields for conductors or insulators are well-known and have a direct intuitive physical interpretation. But, for gravitino fields, or spinor fields in general, the appropriate boundary conditions, describing as they do the effect of a material boundary on these fields, are much less clear. In a supersymmetric theory, the boundary conditions as well as the Lagrangian must be supersymmetric. But in the formulation of the supersymmetric boundary conditions, one can not really make use of the "supersymmetric material properties" of the boundary since it is not clear what these should be. Instead, it is better (reverting really to Faraday's and Maxwell's formulation) to introduce boundaries by additional charges and currents which simulate the expected effect of boundaries. To sketch this method, consider the supersymmetric Lagrangian:

$$L_0 = \frac{1}{2}\left((\partial_\mu \varphi)(\partial^\mu \varphi) - m^2\varphi^2 + \bar{\psi}(i\partial\!\!\!/ - m)\psi\right). \tag{55}$$

The supermultiplet consists of a boson ϕ and a fermion ψ of the same mass m. To introduce boundaries consider, instead of (55), the Lagrangian:

$$L = \frac{1}{2}\left((\partial_\mu \varphi)(\partial^\mu \varphi) - m^2\varphi^2 - j(x)\varphi^2 + \bar{\psi}(i\partial\!\!\!/ - m)\psi - \rho(x)\bar{\psi}\psi\right). \tag{56}$$

In (56), j(x) is a c-number distribution of the boson field; ρ(x) is a similar distribution of the spinor field. These fields j(x) and ρ(x) will eventually be so chosen that their effect on ϕ and ψ simulates boundaries. The requirement of supersymmetry for (50) is introduced by demanding that the supersymmetric variation of the action vanishes. Thus,

$$\delta S = \delta \int L \, d^4x = 0 \ . \tag{57}$$

It can be shown after some calculations that (57) transcribes to

$$j = m\rho \qquad \forall \ x, y, z \ , \tag{58a}$$

$$\partial_\mu \varphi = 0, \ \text{if} \ \rho \neq 0 . \tag{58b}$$

(58) gives the conditions on j and ρ, guaranteeing the supersymmetry. It is possible to calculate the zero-point energies of the fermi and bose fields of the system (56) without imposing the supersymmetry conditions. After a rather lengthy calculation, one finds

$$E_B^{(0)} = \frac{1}{4\pi} \int d\omega \int d^3x \, \log \left[1 + j(\vec{x}) \, G^0(\vec{x} - \vec{x}', \omega) \right]_{x=x'} \tag{59a}$$

$$E_F^{(0)} = -\frac{1}{4\pi} \int d\omega \int d^3x \, \log \left[1 - \rho(x) \, S^0(\vec{x} - \vec{x}', \omega) \right]_{x=x'} . \tag{59b}$$

G^0 and S^0 are the time Fourier tranforms of Green's functions of the field equations of ϕ and ψ obtained from L_0. In general, $E_B{}^0$ and $E_F{}^0$ are not equal. To get more detailed results, consider the case where:

$$j(x) = j_1(x) + j_2(x) = m\rho(x) . \tag{60}$$

Pick, in particular, a configuration where:

$$j_1(x,y,z) \neq 0 \text{ only if } \left(-R_0 - \frac{a}{2}\right) < x < -R_0 + \frac{a}{2}$$

(61a)

$$j_2(x,y,z) \neq 0 \text{ only if } R_0 - \frac{a}{2} < R_0 + \frac{a}{2} .$$

(61b)

Further, $J_1(x)$ and $J_2(x)$, when not zero, are constant (=g); R_0 and a are fixed numbers. The assumption (61) combined with (60) represents a distribution of charges and currents, $\rho(x)$ and $j(x)$, which differs from zero only in two plates of width a, located at $x = -R_0$ and $x = +R_0$ This particular choice of charges and curents (60) and (61) evidently simulates the original Casimir configuration. With the choice (60) and (61a), the expressions for the fermi and bose energy can be further reduced:

$$E_F = -\int d^3k \int_{-\infty}^{+\infty} dx \log\left(1 + 2j(x)\, G^0\right)$$

(62a)

$$E_B = +\int d^3k \int dx \log\left(1 + j(x)\, G^0\right) .$$

(62b)

It is striking that even under these conditions, there is still no cancellation of the fermi and bose contributions. This is because (60) and (61a) do not yet meet all the necessary conditions for supersymmetry. (60) does satisfy (58a), but not the condition (58b) that whenever $\rho \neq 0$, $\partial_\mu \phi = 0$. This condition states that the normal derivative of the bose field is to vanish inside the plates. Thus, the supersymmetry requires a Meissner-type effect for the current of the bose field. No bose current can exist inside the region where there is "supersymmetric" matter. This condition can be met only when g is infinite (supersymmetry requires a perfect conductor). Analysis of (59) and (62) shows that in this limiting case only, all the boundary conditions are satisfied, and then indeed there is no Casimir effect. But only in this special limit does the precise bose-fermi cancellation occur. This simple example illustrates explicitly that a supersymmetric Lagrangian is not sufficient to guarantee the absence of a Casimir effect. Boundary conditions are needed as well.

(4) Conclusions and More Questions

a. Several arguments have been collected here to motivate a further study of Casimir-like or Casimir-related effects in supersymmetric theories. One interesting aspect of this suggestion is that at finite temperature supersymmetry is necessarily broken. This is in contrast to more conventional symmetries which tend to be restored at higher temperatures rather than broken. It was, for example, shown[27)] that for finite temperatures the masses of the A,B and ψ particles in the Wess-Zumino model are different from the T=0 values. They are also different in different temperatures ranges:

$T=0$	$T<T_c$	$T>T_c$
$m_A^2 = m_0^2$	$m_A^2 = m_0^2 - \frac{1}{4}g^2T^2$	$m_A^2 = -\frac{1}{2}m_0^2 + \frac{1}{8}g^2T^2$
$m_B^2 = m_0^2$	$m_B^2 = m_0^2$	$m_B^2 = m_0^2 + \frac{1}{8}g^2T^2$
$m_\psi^2 = m_0^2$	$m_\psi^2 = m_0^2 - \frac{1}{4}g^2T^2$	0

(63)

Consequently, a finite temperature supersymmetric theory will show a Casimir effect. But the magnitude of that effect will generally be small. Because of the many competing factors, it is extremely difficult to assess the net result with any degree of confidence. One could envisage the following type of scenario for an early universe:

Stage 1: The universe is very hot and very small. The high temperature breaks the initial supersymmetry so a Casimir effect is possible, but typically the high temperature Casimir effect is small. On the other hand, for small confinement volumes, the Casimir effect is large; consequently, the net result is unknown.

Stage 2: In this stage the universe cools and at least some dimensions expand. As the supersymmetry gets better, the Casimir effect becomes less effective. As the usual expansion presumably continues, the dimensional reduction slows down.

Possible Stage 3: At this later stage, strict supersymmetry is broken (this time, not as a finite temperature effect, but due to other mechanisms), the temperature will drop, there could be a Casimir effect, and dimensional reduction might resume. Needless to say, this whole scenario is extremely speculative and really without any sound foundation. It was noted only to stress the many ways in which the Casimir effect might operate.

It would seem much more reasonable and less contrived to use the insight gained in section (3c) and use the boundary conditions as the mechanism which break the supersymmetry. It may well be that the boundary conditions required by cosmology automatically break the supersymmetry. This would be a symmetry breaking with a minimum of arbitrariness; it would certainly yield a Casimir effect; it could possibly yield an effect of the right order of magnitude without invoking too many auxiliary fields. It would be interesting to investigate a specific example and numerically calculate the Casimir effect for a system where the cosmological boundary conditions break the supersymmetry.

b. It is probably wise to be extremely careful when invoking phase transition type mechanisms in higher-dimensional spaces. It is well known that the critical exponents of the Ising model in spaces with $d>4$ are the mean field exponents. These higher dimensions appear to obliterate at least a section of the dynamical character of the system. It is further known[28)] that the ϕ^4 field theory is trivial in spaces where $d>5$. This might indicate that spaces of dimension 10,11, etc. cannot sustain phase transitions at all. Thus, the phase transition picture and the physical reality of extra dimensions might not be compatible. Only after compactification has taken place can a genuine phase transition occur. This observation does nothing to clear up an already murky picture. But it does indicate that the problems are subtle and it is unlikely that a single qualitative insight can clarify even the main features.

References

1. T. Kaluza, Sitz. Preuss. Akad. Wiss. Phys. Math. Klassa., 966 (1921).
2. O. Klein, Z. Phys. 37, 895 (1926).
3. E. Cremmer and B. Julia, Nucl. Phys. B103, 399 (1976); J. Scherk and J.H. Schwarz, Phys. Lett. 82B, 60 (1979).
4. A. Pais, Physica XIX, 869 (1953).
5. E. Alvarez and M. Belén Gavela, Phys. Rev. Lett. 51, 931 (1983).
6. S. Barr and L.S. Brown, Univ. of Washington preprint 40048-01P4.
7. E.W. Kolb, comments at this conference.
8. D. Sahdev, Phys. Lett. 137B, 155 (1984); Univ. of Pennsylvania preprint UPR-0248T.
9. H.B.G. Casimir, Proc. Kon. Ned. Akad. Wetenschop., 51, 793 (1948).
10. E. Witten, Nucl. Phys. B186, 412 (1981).
11. M. Dresden, Physica 110A, 1 (1981).

12. A. Guth, Phys. Rev. D23, 347 (1981).
13. A.D. Linde, Phys. Lett. 108B, 389 (1982).
14. A. Albrecht and P.J. Steinhardt, Phys. Rev. Lett. 48, 1220 (1982).
15. S. Weinberg, Gravitation and Cosmology, (Wiley, New York, 1972); see especially pages 375-412.
16. P.G.O. Freund, Nucl. Phys. B209, 146 (1982).
17. Some details of Casimir-type calculations are given in M. Dresden and D. Rohrlich, Physica, in press.
18. See especially T. Appelquist and A. Chodos, Phys. Rev. Lett. 50, 141 (1983); Phys. Rev. D28, 772 (1983).
19. A. Chodos, "An Introduction to Kaluza-Klein Theories", edited by H.C. Lee (World Scientific, Singapore).
20. D. Rohrlich, Phys. Rev. D29, 330 (1984).
21. T. Appelquist, A. Chodos, and E. Meyers, Phys. Lett. 127B, 51 (1983).
22. S. Weinberg, Phys. Lett. 125B, 265 (1983).
23. P. Candelas and S. Weinberg, Part. Phys. B237, 397 (1984).
24. A. Chodos and E. Meyers, Annals of Physics, to be published.
25. A. Chodos, talk given at this conference.
26. A much more detailed description of this material is contained in a forthcoming paper of M. Dresden and J. Zanelli, Physica, in press.
27. L. Girardello, M.T. Grisaru, P. Salomonson, Nucl. Phys. B178, 331 (1981).
28. M. Aizenman, Commun. Math. Phys. 86, 1 (1982).